NORTHERN HEMISPHERE

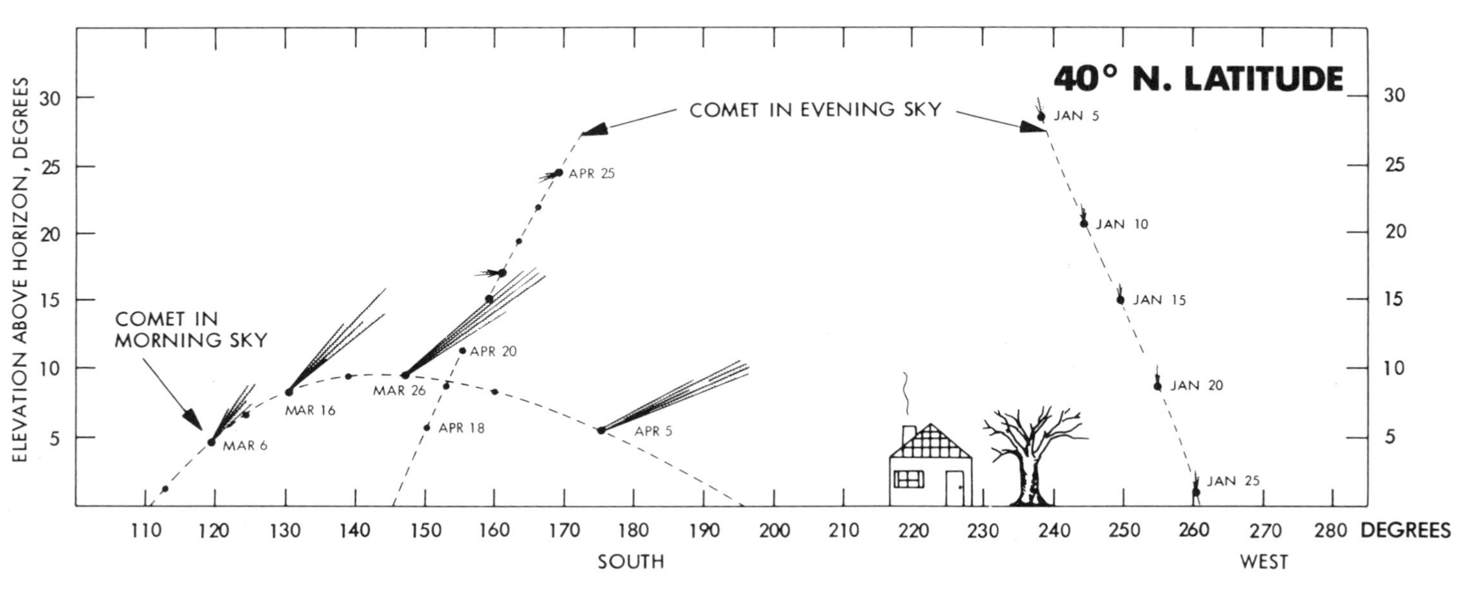

Comet Halley observing conditions in 1986 for observers located at 40° North Latitude (for appropriate countries see Harpur's Guide pages 36 and 37). Comet positions are given for beginning of morning astronomical twilight or end of evening astronomical twilight. Viewing with binoculars and ideal observing conditions are assumed. Astronomical twilight is approximately one hour before Sun rises or sets. For example 5th April one sees Comet low on horizon looking South.

THE OFFICIAL HALLEY'S COMET BOOK

THE OFFICIAL HALLEY'S COMET BOOK

BRIAN HARPUR

HODDER AND STOUGHTON
LONDON SYDNEY AUCKLAND TORONTO

British Cataloguing in Publication Data

Harpur, Brian
The official Halley's Comet book.
1. Halley's comet
I. Title
523.6′4 QB723.H2

ISBN 0 340 36511 0

 Photoset by Rowland Phototypesetting Ltd, Bury St Edmunds, Suffolk. Printed in Great Britain by St Edmundsbury Press, Bury St Edmunds, Suffolk. Hodder & Stoughton Editorial Office: 47 Bedford Square, London, WC1B 3DP.

CONTENTS

Illustration Acknowledgments

The illustrations in this book come from the archives of the Halley's Comet Society, with the exception of the following which are reproduced by permission of the following: *Illustrated London News* (pp 14, 48, 49, 50), the BBC Hulton Picture Library (pp 84, 86, 87, 88, 97, 98, 105, 112, 115), United Features Syndicate Inc (p 134) and the *Evening Standard*, London (p 134).

Introduction

My fascination with Halley and his Comet began in 1928 when I was about ten years old. I picked up a book called *The Story of the Heavens* by Sir Robert Ball, the late nineteenth-century Irish Astronomer Royal, and within its pages I stumbled upon what was – and still is – for me the incredible tale of a galactic gipsy which has swung from the deep freeze of our outer solar system backwards and forwards around our sun every seventy-five to seventy-eight years since 240 BC, and has been computed for doing the same perhaps a further thousand years before that. This majestic metronome of the cosmos beating out its now predictable orbit, thanks to the genius of the amazing Edmond Halley, reminds us on each of its returns of the mysterious order of the universe and the clockwork precision of its rhythms.

For me it is an awesome, humbling and inspiring thought that from the late autumn of 1985 to the spring of 1986 we shall have the opportunity of observing the same Comet as did the ancient Chinese astronomers, as did Nero, St Paul, Attila the Hun, and as Mohammed, William the Conqueror and the unfortunate King Harold, Genghis Khan, Shakespeare, Tennyson and Tolstoy (who probably saw it twice). I know that one shares the same experience with them when seeing the sun and the moon, but one cannot liken such a prosaic daily occurrence to a once in a lifetime phenomenon.

Thus on August 4th, 1975 I registered Halley's Comet Society and Halley's Comet Ltd. The former attracted a wide variety of founder members at my personal invitation on the intriguing basis of having no rules, no committees, and no annual subscriptions. The only obligations were the purchase of a tie (for the men) bearing our special '1986' logo, which I designed with the '9' shaped like a comet, and to pronounce Halley as 'Hawley'. My wife, Mimi, designed a medallion (for the women) bearing the same logo. Halley's Comet Ltd adopted the 1986 logo as 'the hallmark of Comet Halley' and registered it as a trademark covering many categories of products and services for merchandising purposes. My aim here was to use my archival material and marketing know-how to license the 1986 logo for merchandising associated with the Comet's return. After recovering my costs (hopefully) I intend to donate the proceeds to the Duke of Edinburgh's Award Scheme and to the funds of the Saints and Sinners Club of London for distribution at their discretion to the many charities which they support.

With the invaluable help of my youngest son James, who did the research, we have compiled this book as the Society's official layman's guide, the contents of which are fully explained in the first chapter.

I owe personally a great debt of gratitude not only to James (much of whose research forms the basis of the historical sections), but also to my wife, Mimi, who gallantly taught herself to type and to read my bad handwriting so that it could be translated into something more legible. I wish also to thank Mrs Overland, my former secretary, for her help in typing and keeping records over the years, and James and I unite in acknowledging with gratitude the help and encouragement given to us by Dr Patrick Moore and Colin Ronan, who did so much to help me set up and sustain my Society since its inception. Colin Ronan's wonderful book *Edmond Halley: Genius in Eclipse* (published by Macdonald, London, 1970) should be read by everyone, as should Nigel Calder's *The Comet is Coming* (BBC Publications) and *The Return of Halley's Comet* by Patrick Moore and John Mason (Patrick Stephens Ltd), all of which gave us excellent background. I would also like to acknowledge Dr Patrick Moore's kindness in going through the book in its proof stage and making with typical thoroughness a number of most helpful comments. It was Colin Ronan's reference to the distinct possibility of Halley having his name pronounced 'Hawley' which gave rise to the Society's convention in referring to him in that way as a bit of conversational 'one-upmanship'. We also thank, for their great help in supplying information and every possible assistance, Miss Ruth Freitag of the Library of Congress, Washington, DC (without whose command of hundreds of relevant references this book could never have been written), Dr Robert S. Harrington, US Naval Observatory, Washington, DC; Joseph M. Laufer, President, Halley's Comet Watch 1986, New Jersey; Liz Moore, *Illustrated London News* Picture Library; the staff of the Public Record Office; Peter Hingley, Librarian of the Royal Astronomical Society; Robin Gorman, Chairman of the committee organising National Astronomy Week in early November 1985 and to Enid Lake and all the other members of that Committee; Michael Dawes of Taylor & Francis Ltd for permission to quote from *Correspondence and Papers of Edmond Halley* by E. F. MacPike (1937); Dr Kiang of Dunsink Observatory, Dublin; Martin Freeth of the BBC, producer of *The Comet is Coming*, May 1981; Dr David Hughes, University of Sheffield; Dr Stuart Malin and Carole Stott of the Old Royal Observatory, Greenwich; Professor Obayashi of the Institute of Space and Astronautical Science, Japan; Mr Izumi Nakanishi of Dentsu Inc.; Captain Harry Home Cook, Hon. Secretary, Halley's Comet Society; Professor Fred Whipple, Smithsonian Institute; and Dr Donald K. Yeomans of the Jet Propulsion Laboratory, Pasadena in the United States; Mr Christopher Walker of the British Museum; Dr Richard Stephenson of Durham University; Dr David Whitehouse of Mullard Space Laboratory; Dr Zarnecki of Kent University; and in particular for his constant interest and helpfulness in so many of my Halley projects the Astronomer Royal, Professor Graham Smith, Jodrell Bank; and to members of the USA Chapter of Halley's Comet Society namely Mr Don Engebretson, Mr and Mrs Robert Zimmerman, Mr and Mrs Donald Budge, Mr Robert Hart and Mrs Harriett Wyatt, and Dr Edward Proctor and his wife Dr Lois Proctor.

I would also like to record my appreciation of the support, encouragement and advice offered me by Mr Michael Marshall, MP, Chairman of the all-party Committee on Space in the House of Commons. For the same reasons I am indebted to Admiral Sir Raymond Lygo, Managing Director of British Aerospace;

to Mr Hugh Metcalfe, Chief Executive of their Dynamics Group (the prime contractor for the European Space Agency's 'Giotto' mission), to Mr Michael Hird, Mr Terry Bickerton and Mr John Humby among other executives in charge of the group's public affairs and information and to Mr Hugh Manning and Mr Hugh Mooney who checked the facts in relation to the performance of Giotto and of the spacecraft itself; and to Mr R. Jenkins, also of British Aerospace, who helped with ingenious ideas for locating the Comet at different latitudes.

I must also record my deep appreciation to Sir Basil and Lady Lindsay-Fynn for their kindness in giving me privileged private access to their home to view and photograph the historic Samuel Scott painting of the 1759 Comet.

My thanks must also be given to Miss Philippa Toomey for licking a messy first draft of the book into shape and last but not least to my friend Mr Laurence Cotterell who has selected and edited the poetry for Chapters 13 and 14 and whose expertise and devotion to the Halley cause steered me through rough publishing seas into the haven provided by Mr Ion Trewin and the house of Hodder & Stoughton. Finally I should thank the City of London's Lord Mayor (elect) Alderman Allan Davis for his enthusiastic support in appreciation of his term of office coinciding with the year of the Comet; to the Earl of Avon for expediting the restoration work on Edmond Halley's headstone at Greenwich, and to the Dean of Westminster, the Very Reverend Dr Edward Carpenter for enabling Edmond Halley to take his rightful place among his peers by being memorialised in Westminster Abbey in 1986.

Brian Harpur

1

What This Book is All About

'Once in a lifetime' is a commonly used expression. But there is an event which occurs in each generation, every seventy-five to seventy-eight years, which is shared world-wide, which lasts about six months, which is predictable years in advance, which promotes fantasy, fables and frenetic excitement, and at the same time is taken to presage doom, pestilence, plague, disease and even the end of the world.

It is the return of the world's most famous Comet, named after a great but little known Englishman called Halley.

This Comet will be visible from one part of the world or another to the naked eye or with the aid of binoculars from the late autumn of 1985 to the spring of 1986. It will then leave our heavens once again on its great orbit that takes it to the far reaches of our solar system (by about AD 2024) before slowly turning round to be pulled back by our sun. It will not be seen again from earth – at least by the naked eye – until the year AD 2061.

This book, especially written as the official Halley's Comet Society handbook, contains, I hope, everything you would wish to know in simple explicit language about the Comet and Edmond Halley. It is designed to be the complete guide to all the facts, myths and legends which surround the visitations of the Comet ever since its appearance was first recorded in 240 BC, up to and including its thirtieth 'apparition' (as the astronomers call it) in 1985–6.

It will give you the facts about the epoch-making events which have happened on or about each of its returns. In AD 66 it hovered over Jerusalem like a glittering scimitar presaging the fall of the Holy City to Vespasian and his Roman army. It was in our skies again in AD 374 when Italy was invaded by the Huns and again in AD 607 when Mohammed began to preach in Mecca.

Halley's Comet is associated with the best-known date in English, if not European, history when in 1066 William of Normandy saw it as a favourable omen telling him there was a 'kingdom awaiting a king'. Harold on the other hand regarded it, quite rightly, as an ill omen for the defence of his realm.

This book will tell you all about its next appearance, including the five giant space probes being sent up to take a close look so that for the first time it is hoped to find out what a comet is. According to the present theory it is a 'dirty snowball' made up of frozen gases and cosmic dust which starts to thaw and send out a spray of melted particles. These form the traditional tail of the comet as it gets closer to the heat of the sun.

With the help of experts I have compiled a detailed guide to tell you when and where and how to view it. On this occasion observers on the ground are going to be denied the spectacle of the Comet on February 9th, 1986 when it is closest to the sun (the astronomers call that 'perihelion'), because at that time the sun comes precisely between ourselves and the Comet. But for many weeks before it gets to that point and for many weeks afterwards, given the right viewing conditions, it will be capable of being seen by the naked eye or with the aid of ordinary binoculars or in small telescopes at some time or another from most parts of the world (see chart, pages 36–7).

Elsewhere in this book (Chapters 9 and 10) is an account of the life of Edmond Halley, a remarkable man who died in 1742 at the age, considerable for his times, of eighty-six. Without him Sir Isaac Newton might never have written his masterpiece, the *Principia*. Yet when Halley died, this link with the Comet which made his name universally acclaimed and brought him renown in the world of astronomy was not even mentioned.

I have tried to include a multitude of interesting conversational points which will enable you to explain to your friends the scientific, historic and romantic significance of Edmond Halley and his Comet. Did you know that it has been seen already, not just months before its sighting as was the case in 1910, but no less than three years ahead? Such has been the dramatic advance in telescopic technology and techniques in the last seventy years. But let us start with the answer to a basic question – what is a comet?

2

What is a Comet?

Comets, such as the one named after Edmond Halley, are often described as 'hairy stars'. The long tails of luminous particles look like strands of hair as they stream away from the Comet when the sun's warmth starts to melt this frozen 'snowball'. Observers of the big comets over the centuries have left us sketches clearly showing the impression of hair being blown away from the head by the solar wind.

As the comet goes round the sun the 'hair' streaming from it always points away from the sun, just as one's own hair is blown backwards by a conventional hair-dryer. The sun is in effect acting as a powerful solar hair-dryer sending out warm currents which melt the frozen tresses on the head of the approaching comet, causing them to ripple and sparkle as they unfurl in a cascade of glory, millions of miles in length.

That is how comets got their name. The word comet comes from the Greek *kometes* meaning long-haired. The word 'perihelion', used by astronomers to describe the exact time when the comet is closest to the sun, comes from the Greek *peri* meaning close and *helios* meaning sun.

Although comets are not at all rare and many less well-known ones visit us at irregular intervals just like Halley's, many people think that a comet goes streaking across the night sky in a blinding flash, never to be seen again. Actually this is to confuse comets with shooting stars (or meteors), which are dust particles that burn up when hitting the protective shield of the earth's atmosphere. A comet can be visible for weeks or months at a time, while hardly appearing to move. Some comets, however, are only seen as faint milky-white patches of light while others can be astonishing spectacles stretching across the sky.

Up to now about 1,000 different comets have been registered and as many as ten new discoveries have been made in a year, mainly as a result of the growing number and enthusiasm of amateur astronomers. So why are five expensive space probes (from Europe, Japan and Russia) being concentrated on Halley's Comet when there are so many other comets around?

Theoretically, new comets, which are the most active and bright, are the best to work on, especially a comet which has not approached the sun before. To send up a satellite from earth to intercept one with a reasonable degree of accuracy one has got to know in advance the precise path and behaviour of the approaching comet. This cannot be properly predicted until the comet has completed a number of its periodic orbits. Comets which come round at short-term intervals are, however, much less bright than new ones and produce much smoother tails, having far less gas and dust.

Halley's Comet is the only one we know of with a well-defined orbit, which still has as much gas and dust to throw out as a new comet.

An accurate representation of the astonishing fan-shaped tail of the 1910 Comet as drawn by Charles Wyllie after viewing it through the large telescope in the Old Royal Observatory at Greenwich.

Comets spend almost all their time in a 'deep freeze' at about 4,675,000,000,000 miles away from the sun. Because it is impossible to describe distances of that magnitude, astronomers have reduced it to just 50,000 astronomical units (A.U.). If this is meaningless, let me add that one A.U. is 149,000,000 kilometres or 93 million miles, being the average distance of the earth from the sun.

Thanks to their isolated frozen state, comets have preserved the origins of our solar system. Their chemical composition has changed very little since its formation, computed to be some 4.6 billion years ago. Scientists are still, however, unsure of their exact composition. A typical comet is thought to consist of three basic parts: a nucleus, a coma and a tail.

The nucleus is the enigmatic part of the comet and cannot always be detected. According to the current, most popular view it is made up of particles packed together with certain frozen substances and is only a few miles in diameter.

The coma (the word, like 'comet', is derived from the Greek for 'hair') is the head or 'halo' of the comet, a great gaseous envelope surrounding the nucleus. It is formed by the sun's heat acting on the nucleus which it melts, releasing various types of particles and volumes of gas. The coma can vary in size. It is usually anything up to 100,000 miles in diameter but can extend to over a million miles, as in the case of the comet of 1811.

The tail can stretch up to tens of millions of miles across the sky. The great comet of 1843 had a tail that was 200 million miles long. Others may appear to have no tails at all, or at best unspectacular ones.

The tail develops only as the comet gets nearer to the sun. Dust particles and ionised gas molecules are driven out of the coma by solar radiation and by solar wind (a continuous stream of atomic particles pouring out from the sun). If the tail is composed of gas molecules, it is straight and fluorescent, i.e. the molecules absorb sunlight at one wavelength and emit it at a different wavelength. If the tail is composed of dust, it is usually curved and its brightness comes from reflected sunlight. Some comets have more than one tail. Donati's Comet in 1858 had a spectacular dust tail with two gas tails that were thin and straight.

Because a tail is created by solar radiation and solar wind, it will always point more or less away from the sun. When a comet has orbited the sun and is on its outward journey, its tail will stream out before it.

The mass of a comet is very low and it receives a blasting from the sun every time it goes round it. Consequently, a comet loses a great deal of material every time it orbits the sun and has a short life-span, at least in comparison with other celestial bodies.

As well as the nucleus, coma and tail, comets are enveloped by a huge cloud principally composed of hydrogen. This cloud is invisible and cannot be detected except by means of a space probe.

The orbits of comets

According to the Ancient Greek astronomical tradition that culminated with Ptolemy, the orbits of the planets were circular. This was in accordance with the Greeks regarding the circle as the perfect figure.

It was not until the seventeenth century that Johannes Kepler showed that planets move on elliptical paths and not circular ones. Sir Isaac Newton developed the work Kepler had done and arrived at the conclusion that any celestial sphere which moves as a result of the sun's gravitational pull on it must have an orbital path that has the shape of a circle, an ellipse, a parabola or hyperbola (see Chapter 16 for definitions).

In practice, the chances of a comet having a circular or parabolic path are extremely slim. Almost all comets have elliptical paths of varying degrees of elongation or eccentricity. If an elliptical orbit is very elongated, it is difficult to distinguish it from a parabolic path when the celestial body in question is near the sun. In fact, when Halley scrutinised the orbit of the Comet of 1682, he thought at first that it had a parabolic orbit. But when he saw a resemblance between this Comet and the one of 1531, he considered the possibility that it might have an elliptical orbit, and the more he worked the more convinced he became. Finally, he was able to predict that the 1682 Comet would return in 1758, and he was right.

If a comet does have an elliptical orbit, it will keep returning to the sun until it disintegrates or is dragged off course on to another sort of orbit by a giant planet. This is because the ellipse, like the circle, is a closed figure. On the other hand, the parabola and hyperbola are open figures and so any comet on a parabolic or hyperbolic path would go round the sun, never to return.

The mathematics involved in calculating the orbit of a comet are extremely complex even today. The calculations are made more difficult by the fact that a comet's path can be altered by the gravitational pull of a giant planet, such as Jupiter or Saturn. This is why the period of time Halley's Comet takes to return to the sun is slightly different every time it comes round to perihelion.

Comets that keep returning to the sun are called 'periodic'. The time taken to complete an orbit varies from comet to comet. Some, called 'short-period' comets, can take from a couple of years to a couple of hundred years. Others, called 'long-period' comets, can take tens of thousands of years. Comet Kohoutek, for example, which came in 1973, will return to the sun in about 75,000 years' time. On the other hand, Comet Encke takes just over three years to complete one orbit.

The origin of comets

Johannes Kepler once said that there are as many comets as fishes in the sea. Indeed, as the years go by more and more comets are discovered. But where do all these comets come from? The answer is still not known for sure but different theories have been put forward.

It has been suggested that comets come from beyond the solar system in the regions amongst the stars beyond. Then, when a comet occasionally strays towards the inner solar system, it is sucked in by the gravitational field of a giant planet like Jupiter. Consequently it is sent on an elliptical orbit towards the sun which it afterwards orbits as a member of the solar system. However, most astronomers now think that comets do not come from beyond the solar system.

There is another theory, first put forward by the French astronomer Lagrange at the start of the nineteenth century, that says that comets are propelled outwards from the giant planets. For example, Jupiter was supposed to pump out comets from an enormous volcano that was identified with the red spot that has been observed on its surface. In the 1950s, a Soviet astronomer called Vsekhsvyatsky answered the objection that the force needed to propel a comet from the surface of Jupiter was too great for the theory to be feasible by proposing that the comets could be shot out from one of Jupiter's four satellites. However, this theory is still thought to be implausible.

The theory that is currently most favoured involves what is known as 'Oort's cloud'. This cloud, according to the Dutch astronomer, Jan Oort, was formed at the same time as the birth of our sun and planets. It lies at the fringe of the solar system and contains millions or even billions of comets. These comets have incredibly slow circular orbits that are constant until they chance to meet a passing star. Then, the perturbations caused by such a star would be enough to send a comet on a path towards the sun. Having reached perihelion the comet would then head off back to its remote home unless trapped into a short-period orbit by the gravitational field of a planet.

The Dirty Snowball and the Flying Sandbank

Although the precise structure of a comet is not known, theories as to their make-up are not lacking and the two most favoured models of cometary structure are called informally the Dirty Snowball and the Flying Sandbank.

The Flying Sandbank model of a comet has been championed by the British astronomer R. A. Lyttleton. According to this model, there is no difference of quality between a comet's nucleus and its enveloping coma. Both are composed of particles of dust that become more concentrated towards the centre, thus giving the false impression of a solid nucleus. As the comet approaches the sun, the interaction of the solar wind with the gases inside the particles causes a tail to emerge and stream off in a direction away from the sun. The closer the comet gets to the sun, the more the dust particles (which are all moving independently) compress together and are ground to finer particles in the resultant collisions.

The Flying Sandbank is not as favoured currently as the Dirty Snowball model developed by the American astronomer F. L. Whipple. This model does have a solid nucleus, which consists of pieces of rocky material packed together with various iced substances such as frozen water, ammonia and methane. As the comet gets nearer to the sun, the solar heat acts upon the ices which, starting with methane, begin to evaporate, producing material that will be driven out by the solar wind, becoming the tail. But this model still leaves something unaccounted for, notably that in some comets a nucleus cannot be detected at all.

Naming of comets

A comet is named after its discoverer or sometimes after the person who first computed its orbit, as in the case of Halley.

A symptom of the popular panic in 1910, when it was thought that the earth would be poisoned by the Comet's tail, were the doom-laden fantasies of artists. This drawing, from a German gardening magazine, shows mushrooms and flowering plants bursting through greenhouses while a Comet fills the sky.

When more than one person discovers the same comet, the first three to register their discovery with the International Astronomical Union have their names attached to the comet in question, e.g. Comet Tago-Sato-Kosaka, whose discoverers were all Japanese.

It often happens that more than one comet is discovered in the same year. In this case, to distinguish one comet from another, a lower case letter of the alphabet starting with *a* is placed after the relevant year to denote the order of discovery. If three comets are discovered in 1988, they will be designated 1988*a*, 1988*b* and 1988*c*. Afterwards, when their orbits have been properly computed, they are permanently given a Roman numeral in order of their reaching perihelion, so that the first to reach it becomes 1988 I, the second 1988 II, and so on.

Brightness of comets

The reason we can see the stars in the sky is because they are self-luminous bodies, i.e. they generate their own light. Planets, on the other hand, reflect sunlight and are not self-luminous.

Comets, like the planets, reflect sunlight and the closer a comet gets to the sun the brighter it becomes. However, it is now known that comets also have the property of fluorescence, i.e. the gas molecules in the coma absorb sunlight at one wavelength and then emit it at another wavelength.

Much information about the chemical make-up of comets comes from spectroscopy, the science of studying the spectra of celestial sources, such as stars and comets, in order to find out what the sources are composed of.

In the spectrum of a comet it is possible to detect evidence for reflected sunlight and also fluorescence from certain gases such as methane, carbon monoxide and cyanogen.

It was the known presence of cyanogen, a toxic gas, that caused some popular panic when Halley's Comet last appeared in 1910, when the earth actually passed through its tail. However, because the gases in a comet's tail are so thin, there was never any danger of poisoning.

It could be that the appearance of Halley's Comet this time will be historic also in respect of its brightness. By September 1984 telescopic surveillance of the comet while it was still very far away revealed marked changes in the light it was throwing out. In the same month Fred L. Whipple, probably one of the greatest authorities on comets, wrote a postscript in a letter to me from the Astrophysical Observatory, Cambridge, Massachusetts, in which he said, 'The variations in the brightness of the still very faint Halley's Comet are a real puzzle. I know of no good explanation for them yet.'

It seems quite certain that Halley's Comet has a number of surprises for us, awaiting explanations which, hopefully, our five space missions intercepting it in March 1986 will provide, if not before.

3

The Search for the Return of the 1986 Comet and the Plans to Track and Intercept It

On the night of October 15th/16th, 1982 a team of astronomers in the United States led by a British graduate student David C. Jewitt, aged twenty-four, with D. Schneider and A. Dressler were the first to detect the return of Halley's Comet a billion miles out in space as it accelerated towards the sun to register its thirtieth recorded reappearance since 240 BC. They had been waiting to sight it ever since November 1977, using special equipment in conjunction with the huge 200-inch telescope on top of the Palomar mountain in California.

This breakthrough put an end to the feverish impatience of astronomers all over the world as they sweated over their observations hoping to go into the history books as the first to spot the return of the world's most famous Comet. Three of them who spent four nights in February 1982 using the 107-inch telescope at McDonald University in the United States combined professionalism and concentration to such a degree they must have come pretty close to seizing the honours, but the Comet just eluded them.

Two others, Michael Belton and Harvey Butcher, who had been working at intervals for several years at Kitt Peak National Observatory, USA, gained some satisfaction from predicting as a result of analysing their frustrations that the Comet might be found by late September or early October 1982. In the end they were only a few days out.

The first photograph then taken shows the Comet as a tiny blurred speck of light surrounded by hundreds of brighter stars and one has to take the astronomers' word for it that this *is* the Comet, because it has turned up in almost the exact spot where all the calculations had indicated it should be found.

The 'watch' for Halley's Comet was set up in a number of ways, involving both amateurs and professionals.

A Halley's Comet Watch was inaugurated by one enterprising amateur, Joseph Laufer, to gather and disseminate information on all aspects of the Comet by means of a periodic newsletter first published in 1982. Mr Laufer of New Jersey, USA, said in an interview he gave at the time: 'I want to be part of the excitement, part of the moment in history . . . the comet reaffirms my faith . . . The fact that there is this wondrous thing out there that keeps coming back like clockwork through a natural predictable process continues to fascinate me, and to remind me of the order in the universe which contrasts so dramatically with an observably chaotic world. For that reassurance alone, I am grateful.'

However, Mr Laufer's splendid initiative must not be confused with the setting up of the official International Halley Watch to secure the co-operation of

This tangled assembly of mirrors and cross pieces is far removed from one's concept of the conventional telescope. The 'Keck' apparatus enables the mirrors to be moved together in different combinations so as to provide, when integrated, the most powerful telescope ever devised, graphically summarised by the claim that it would be able to pick up the light of one single candle on the moon.

professional and serious amateur astronomers and government agencies all over the world. Its main centre where observations and data are being co-ordinated and analysed is the National Aeronautics and Space Administration's (NASA) Jet Propulsion Laboratory in Pasadena, California. The most famous astronomers and scientists are joining with universities, astronomical societies and giant industrial companies in nearly fifty countries covering all five continents in conducting what has become an historic and unprecedented level of international co-operation.

The name to conjure with here is Dr Donald K. Yeomans of NASA. He is a world authority on the comet and the calculations pinpointing its 1985–6 orbit. By June 1983 he had noted that since the first sighting in October 1982 there had been no less than fourteen further sightings from the great telescopes sited in Arizona, Hawaii and Chile, as well as Mt Palomar. The fact to remember about telescopes is that the great magnification they bring is influenced by the amount of pure light their mirrors reflect. If the light is diffused or weakened in any way by there being a full moon or the glow of a city's street lighting, for example, then it is not so effective. That is why the large telescopes are sited in remote dark places and as high up as possible to escape the distortions of the earth's atmosphere at ground level.

Nevertheless there are thousands of small telescopes at work, mostly used by dedicated part-time or amateur astronomers, in back gardens and in small observatories accessible to the public, which have played a major role in checking and adding to astronomical data. This is particularly the case in discovering new comets. They form an important part of the vast network set up by the official International Halley Watch.

Also playing a major role will be the Royal Greenwich Observatory, now based in Herstmonceux in Sussex, England. They have built, with the co-operation of the Spanish government, a complex to accommodate three telescopes on the top of La Palma, some 2,400 metres above sea level in the Canary Islands. This includes the computer-operated Isaac Newton telescope. In 1986 it will be joined by the even bigger William Herschel, measuring 4.3 metres (165 in), which was built in Newcastle-upon-Tyne, England.

Although La Palma is 2,000 miles away, British astronomers at their headquarters in Sussex will carry out their observations without leaving the country, or even their armchairs. Using telephone lines as their links, they will instruct the telescopes which way to move, check what they pick up on a TV screen, and even change the photographic plates and filters, while collecting all kinds of data. There are five telephone lines available, and if they become too congested with the flow of information, recordings on magnetic tape can be flown back the next day.

An even more dramatic move is the American plan for 1986 to put the world's most expensive telescope, costing about £300m, into orbit, well clear of any distortions of the earth's atmosphere. The instrument will be shot into space aboard America's very successful Shuttle spacecraft. There is even talk of turning the Shuttle's 150-foot-long external fuel tank, which weighs about 30 tons, into an instrument to detect gamma rays. At present this tank has a useful life of less than ten minutes. It carries fuel to assist with the blast-off and, as soon as the Shuttle starts to leave the earth's atmosphere, it is jettisoned.

Should there be a launch of a telescope into space when Halley's Comet leaves our skies in 1986, it might well be possible for the first time to observe the Comet all the way to the point of its slow-speed turn-round on the outer regions of our solar system, in about thirty-seven years' time, and then to watch it gradually accelerating on its way back to visit us again in 2061. In this event the 1986 apparition will be truly historic in that Halley's Comet will never again be 'recovered'. This is the astronomers' word for first detecting the return of a 'lost' comet which comes back at periodic intervals.

But the story of the development of bigger and better telescopes shows no end. By 1992 astronomers expect to have the use of the world's largest optical telescope sited on top of Mauna Kea, an extinct volcano nearly 14,000 feet high, in Hawaii. This gigantic project, called Keck after the Foundation which has put up millions of dollars to fund it, will have a new kind of mirror system consisting of a computer-controlled complex of thirty-six six-sided pieces of special glass. The pieces, each six feet wide and three inches thick, will be able to interlock and move in unison to become integrated as a mirror nearly 400 inches in diameter. This will have twice the width and four times the light-gathering capacity of the Hale apparatus at Mount Palomar. This new design helps to overcome many of the construction problems of conventional telescopes and is said to 'push back the visible limits of the universe by *billions* of light-years' (*Time* magazine, January 21st, 1985). When one recalls that a light-year is 186,000 miles a second × 60 seconds a minute × 60 minutes a hour × 24 hours a day × 365 days a year it is a relief to hear the simple remark by Mr Howard Keck, who put such mind-boggling stuff into perspective when he said: 'I'm told it will permit one to see the light of a single candle from the distance of the Moon.'

In addition to enormous ground-based observations there are spacecraft being sent up by Russia, other European countries and Japan.

Despite being observed for thousands of years, very little is known about comets and an encounter with Halley's is one of the last great exploratory challenges which can unlock the secrets of our solar system and where major surprises can be expected. This is why Sir Bernard Lovell, the great British astronomer, said, 'Giotto [the European Space Agency Mission] is one of the few visionary and exciting projects of this decade.'

The probes being made by the Soviet Union followed by Japan and then – the major one – by the European Space Agency will, it is hoped, supply many of the answers. The launch of the European Space Agency satellite involving a ten-nation team led by British Aerospace was called Giotto, after Giotto di Bondone, the painter of Florence. He saw Halley's Comet in AD 1301 and painted it into his interpretation of the Nativity as the Star of Bethlehem when decorating the interior of the Arena Chapel in Padua, Italy. His realistic impression might be called the first scientific portrayal of Halley's Comet recorded in history.

The Comet is travelling at enormous speed in the opposite direction when the encounter with Giotto takes place in March 1986. They will pass each other with such velocity that a microscopic dust particle coming from the Comet's nucleus can hit the spacecraft fifty times faster than a bullet from a gun. This means that something weighing as little as one tenth of a gram, or 0.0035 of an ounce, can

The Great "Giotto" Encounter — at a glance

Relative positions of Earth, Comet, and "Giotto"
(using information supplied by courtesy of Dynamics Group British Aerospace — Prime Contractor for the ten country team Giotto Project).

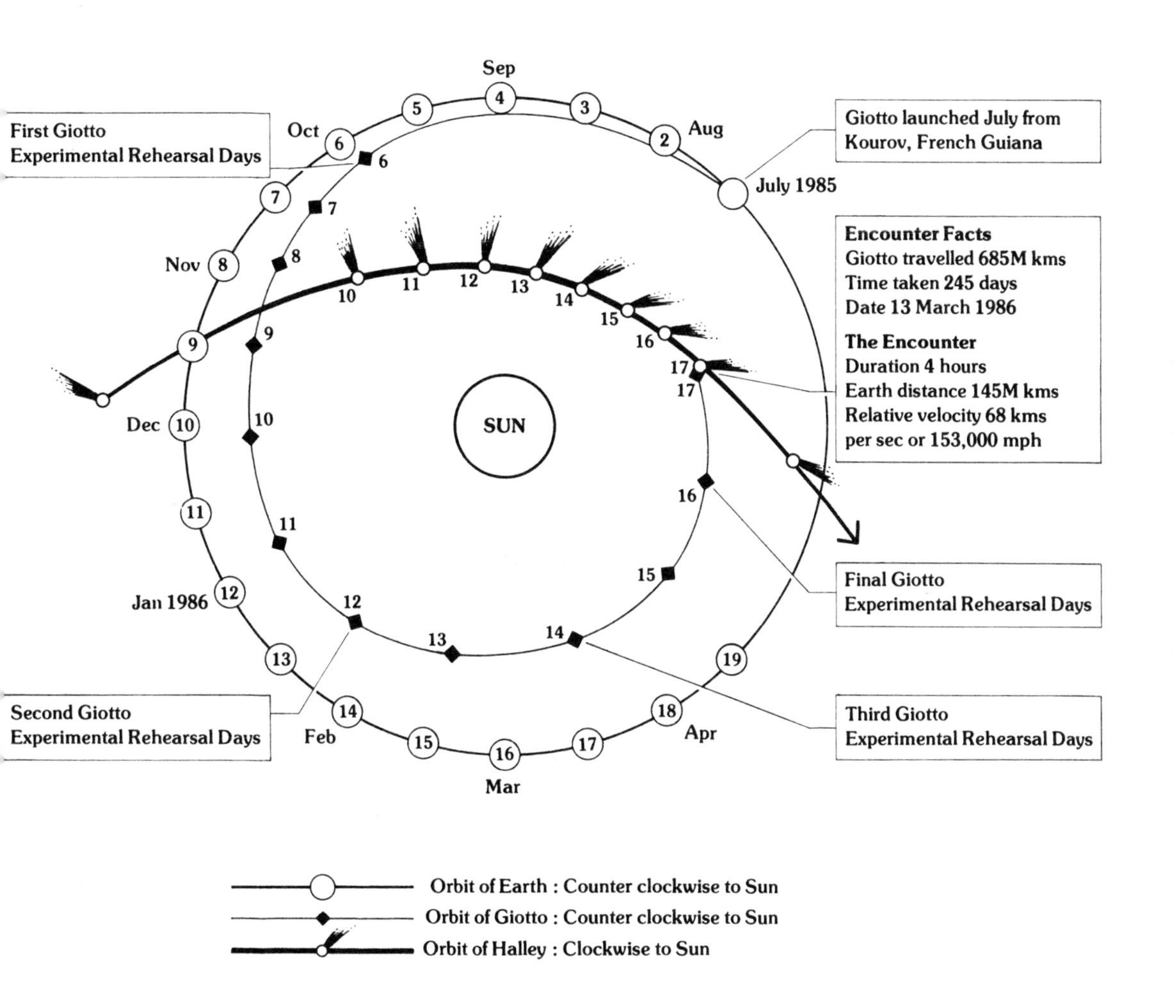

Orbit of Earth : Counter clockwise to Sun
Orbit of Giotto : Counter clockwise to Sun
Orbit of Halley : Clockwise to Sun

How to plot the relative positions:

The orbit of each is numbered to correspond with the dates given for each orbit. The even numbers indicate the start of each month, the odd numbers a mid month date. Examples — December 1985 at Earth (10) Giotto at 10◆ is just drawing ahead and moving inwards slightly towards the Sun, while Comet at 10 is going the other way towards its Perihelion in February (positions 14 and 15). In mid March 1986 the Earth is at (17) and both Giotto 17◆ and Comet 17 meet.

whizz through an aluminium barrier eight centimetres or three inches thick. It needs a shield of some sort but to provide even an ineffective one in the shape of a single sheet of eight centimetres of aluminium would impose a prohibitive weight penalty at lift-off of about 600 kilogrammes or nearly 1,300 lbs.

The simple solution is to put a couple of bumpers in front of the space vehicle. The front one is paper thin, one millimetre or 0.039 inches, and the second one, about twenty-five centimetres (10 in) behind it, is a little thicker. Should a dust particle hit the front flimsy, a terrific amount of energy would be released, fragmenting and vapourising the dust particle. The cloud it forms expands into the space between the two bumpers, and by the time it strikes the second one it is so dissipated over a much larger area that the impact can be largely absorbed. This technique reduces the weight of the shield from 600 to sixty kilogrammes, a very considerable reduction.

In June 1984 five astronomers from the south of England studied the dust particles round the Comet from a volcanic peak in Tenerife, obtaining excellent results from 8,000 feet in skies of exceptional clarity. The leader of the expedition, John Green, said that the Comet was surrounded by countless icy particles ranging in size from a grain of sand to a football. The aim of the expedition was to establish the chance of Giotto surviving its hazardous journey.

Although the size of the Comet's coma, or 'halo', is somewhat greater than the distance from the earth to the moon, Giotto's very high speed in relation to the Comet means the duration of the encounter will be only some four hours. Because there is a real possibility that Giotto will not survive its passage through the coma unscathed, all the data collected by the scientific instruments on board, which includes a colour television camera, will be transmitted immediately to earth.

This creates its own problems, however; the antenna which beams the radio signals must be pointed precisely to earth throughout the entire encounter. If the antenna is seriously disturbed by the spacecraft colliding with a large particle some loss of data will inevitably result while the spacecraft attempts to recover the situation. With a sufficiently large impact, however, the spacecraft may be unable to regain the correct pointing angle, but calculations indicate that this possibility only exists to any significant extent during the last few seconds of the spacecraft's approach to the nucleus of the Comet.

The Giotto spacecraft, designed to carry complicated electronic gear and equipment (including camera and spare batteries), is only three metres (about ten feet) from the tip of its antenna to its bumpers, only a little over half that distance in diameter, and by the time of the encounter (when it has exhausted its solid propellant fuel) will only weigh 430 kilogrammes or 900 lbs. It is a feat of design and compact packing comparable to the entire contents of Buckingham Palace and the White House together being carried away in a single suitcase.

Scientists will also have had to take into account that Halley's Comet, like other comets, does not go around the sun on the same plane or level as that which the earth and the other planets roughly conform to, so, as it shoots across the main solar system, it will come in at a different angle and cut across our orbit going from south to north in November 1985. Then, on its way out, it crosses our orbit again from north to south in mid-March 1986.

It is essential for Giotto to fix its rendezvous with the Comet around either of these times when it is closest to the path taken by the earth because Giotto has to conserve all its energy for lifting the greatest possible scientific pay-load.

It has been decided to try and make the encounter about March 13th, 1986, because it is then that the Comet is most active, having just passed closest to the sun a month before, and it is also roughly equidistant from both the sun and the earth, a distance of 81 million and 94 million miles respectively. The difference of 13 million between the two is the equivalent of give or take an inch in astronomical terms.

All the most significant data will be secured as Giotto flies by the nucleus at a velocity of over 150,000 miles an hour, during the four hours of the actual encounter.

The heat of the sun causes the gas and dust to swirl off the Comet in an unpredictable way, constantly creating calculating problems. In addition, the nucleus cannot be seen from earth, and its position inside the coma can only be guessed at with an accuracy of 500 kilometres or about 300 miles. It is therefore highly unlikely that there would be a head-on collision of spacecraft and Comet.

By the middle of March 1986, when it is hoped that Giotto will be sending back colour pictures of the Comet, the earth will still be trundling along behind on a slightly wider orbit. This means that the earth will get closest to the Comet a few weeks after the space mission has been completed. It will be on April 11th, 1986 when the distance separating them is the smallest (and when the best view of the Comet can probably be obtained from the southern hemisphere) i.e. 59,000,000 kilometres or 38,469 million miles.

All the space missions have been planned on the assumption that the Comet will have a nucleus of frozen ice (the 'dirty snowball' theory). They will be totally unable to re-adapt themselves if, for instance, the nucleus turns out to be a whirling storm of hailstones, which some scientists envisage as a possibility. In this event Giotto, for instance, might be shattered before its camera can click one shutter or its radio send out one 'Mayday' bleep.

No fewer than twenty-one contractors are being co-ordinated by the Dynamics Group of British Aerospace (the prime contractors) from ten different countries – Germany, Denmark, Belgium, Italy, Netherlands, France, Austria, Switzerland, Sweden and the United Kingdom. Giotto will be controlled from the European Space Agency's operations centre at Darmstadt, West Germany, and the experimental data it relays will be received via the sixty-four-metre dish antenna of the Parkes Radio Telescope in Australia.

There will be four other space missions to meet the Comet, two from Russia called Vega I and Vega II and two from Japan, MS-T5 later named Sakigake (pathfinder/explorer) and Planet A. These four spacecraft will each make a high-speed 'fly-by' of Halley's coma at different angles and, when their data has been co-ordinated, it is hoped that a three dimensional picture of the nature of the nucleus of the Comet can be put together.

The Japanese, with their innovative and brilliant engineering and electronic skills, are providing exceptional space missions. When I met Professor Obayashi in March 1984 at Tokyo's Institute of Space and Astronomical Science, he showed me the MS-T5 spacecraft, and predicted that it might be launched about 4 a.m.

on January 4th or 5th, 1985. (In the event it was launched on January 8th.) The MS-T5 will also help to analyse the solar wind and, as it will precede Planet A, scheduled for launching in August 1985, the information it sends back will benefit the performance of Planet A and the Giotto missions.

While the Giotto and Russian space projects can be described as self-destruct ('kamikaze') type missions because they are likely to be obliterated by the particles as they close in on the Comet's nucleus, the Japanese Planet A will skirt the nucleus at a much safer distance.

Giotto is expected to get within 500 kilometres of the Comet, and it is estimated that Planet A will pass it within some 10,000 kilometres on March 8th, 1986. Planet A will therefore be twenty times further away and will be equipped with a camera especially designed to pick up the ultra-violet radiation from the Comet.

The Japanese satellite weighs only 135 kilos (about 300 lbs) and has a diameter of 1.4 metres (about four and a half feet) and is a cylinder of only seventy centimetres (under three feet) long. It is roughly three times smaller than the Giotto satellite, but of course does not have so many functions to perform. The Japanese have launched many satellites to orbit the earth, but this is the first time that they will have organised an artificial planet, tiny though it may be, to orbit the sun.

The Russian Vega mission is divided into two parts. Instead of direct targeting, like Giotto, MS-T5 and Planet A, the Russians are aiming to carry out probes of Venus, and to drop capsules and balloons into its forbidding atmosphere. They will then use the accelerating gravitational pull of Venus to deflect the two Vega satellites into the path of Halley's Comet. Venus, in effect, will act as a booster. Its orbit of the sun on this occasion produces a convenient time for the encounter with Halley to be made possible.

The first Vega, launched in December 1984, reached Venus in June 1985, and was recycled with the aim of being the first to reach Halley on or about March 6th, 1986, and pass it at a distance of about 6,000 miles. The second Vega, also launched in December 1984, is a duplicate of the first, and will home in on Halley after Vega I on or about March 9th. It acts, in a sense, as a 'fail-safe' measure, because Vega I may be destroyed by the Comet's high-speed dust particles if it gets too close. Consequently, Vega II will be programmed to fly by at a safe distance. If Vega I comes through unscathed, then Vega II will be pushed closer, but as its fly-by speed is even faster than Giotto's, over 170,000 mph, it cannot narrow the distance by more than two or three thousand miles, compared with Vega I. Were it to do this, it would not be able to get satisfactory results because its velocity would not give it enough time.

NASA, the United States space agency, was greatly disappointed not to be able – as a result of its budget cut in 1981 – to participate in intercepting the Comet with a custom-built probe of its own. But this has not prevented the United States from being involved. As well as being the home of the International Halley Watch, the United States already has two craft in space making experiments entirely unrelated to the Comet, but which will now be used to gather some invaluable information about it.

It so happens that the planet Venus and the Comet will pass within about 40,000,000 kilometres of each other on February 4th, 1986, just five days before

There are few contemporary representations of Edmond Halley. This portrait was commissioned from D. Janvrin by Halley's Comet Society in 1980.

Right: *The Adoration of the Magi* by Giotto de Bondone with the comet of 1301 which he used as a model for the *Star of Bethlehem* seen at the top. It can be viewed in the Arena Chapel, Padua, Italy.

Below: Halley's Comet in 1759. It was painted by Samuel Scott at that time and is significant not only because of the nicely observed figures and landscape of that period, but also because this was the comet whose appearance Halley correctly predicted, although he himself did not live to see it.

the Comet reaches its closest point to the sun. Thus the American Pioneer Venus orbiter, which has been circling Venus since late 1978, will have the stage all to itself because the other spacecraft will not meet the Comet until a month later. The Pioneer orbiter has equipment which will be used to measure some of the chemical properties of the Comet and also record any irregular behaviour of the Comet caused by an increase in heat as it gets closer to the sun.

Finally no reference to the exciting plans to intercept the Comet by space vehicles designed for that purpose should omit the link they have with the dramatic and historic event taking place on September 11th, 1985: the first ever interception by a spacecraft which was never originally intended for such a role. This is the date that another comet, Giacobini-Zinner, passes the earth's orbital plane and this is how it came about.

The spacecraft involved was launched by NASA in 1978 with the object of monitoring the solar wind before it strikes the earth's magnetic field. Originally named ISEE-3 (short for International Sun-Earth Explorer 3), it was recalled much closer to home in December 1983 to check the moon's effect on the earth's magnetic field.

It has now been pushed out into deep space once more and renamed the International Comet Explorer (ICE for short, hence it may still be pronounced ICEY, as it was before). In its present course it will meet Giacobini-Zinner in what will be a close encounter of the first kind.

ICE was not designed to study comets, but it will pass through the tail of this comet about 10,000 kilometres downstream from the nucleus, thereby picking up in the process some clues about plasmas and magnetic fields. But, most important, it will be able to sample the solar wind in the path of the incoming Halley's Comet, consequently making a useful contribution to the Halley studies. The International Halley Watch so believes in the importance of the ICE research that it is giving it much the same attention as to Halley's Comet itself.

Perhaps the most marvellous thing about all this activity is that Halley's Comet has become a focal point for world unity. Fifty different countries are forgetting their nationalism and political differences to participate under the International Halley Watch in the free and liberal exchange of research and general information. As I write, a *Sunday Times* headline over an article detailing the friendly planning taking place between the Eastern and Western blocs and Third World countries crystallises it in just five words: 'HALLEY'S COMET THAWS COLD WAR'. Dr Donald K. Yeomans has summed up this spirit in one simple but dramatic sentence: 'It is ironic that the same Comet that has caused fear and misunderstanding throughout history should now serve as a focal point for an unprecedented level of international scientific co-operation.'

Halley's Comet Watch

Diagram showing approximate Comet position every 10 days Nov 1985 – Apr 1986

NB speed of Comet is given in MPH for each Comet position.
Approaching Sun it gets faster. Departing it gets slower.

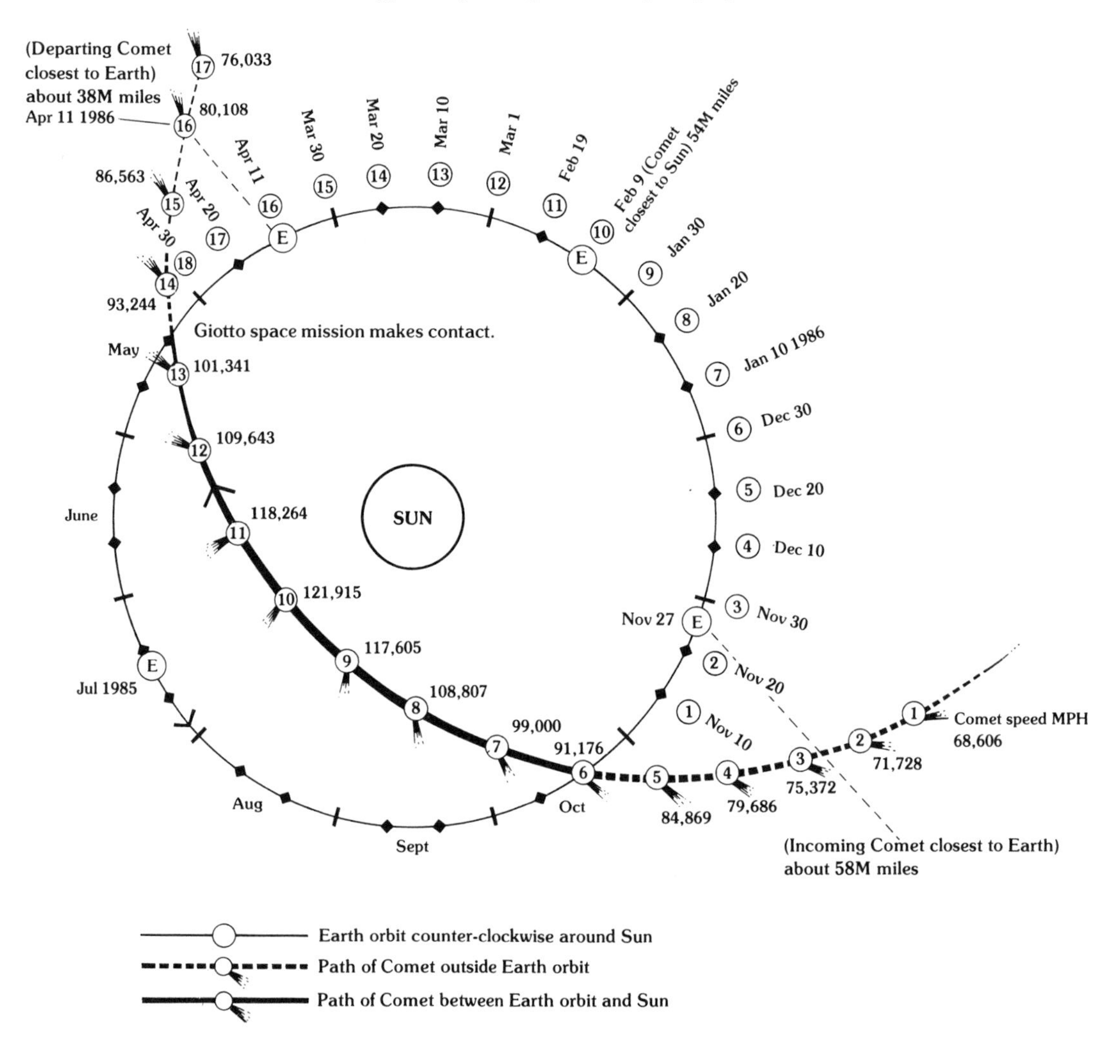

Directions

Check the date on earth orbit. Against it you will find a number. Then find the corresponding number on the Comet's path. That gives your respective positions (approx.) at the date in question.
(Example: On December 30th (position 6) you will see that the Comet's path and earth's orbit almost coincide. But by this time the earth has moved further away from its November 27th position and the Comet is moving closer to the sun in the opposite direction.)

4

Where, When and How to See Halley's Comet

It is a complicated business working out how and where to see Halley's Comet. Just consider some of the more simple facts. The earth is rotating and completes its 360° revolution in twenty-four hours. While doing this it is orbiting round the sun over a period of 365 days – actually a few hours more, which explains why we have a leap year every four years. The speed of the earth as it goes on its 584-million-mile journey around the sun is 66,600 miles per hour.

Meanwhile the Comet is hurtling into our skies at ever increasing speed as the sun's pulling power exerts its growing attraction and whipping around it in the opposite direction to the orbit of the earth. At that point it will be doing about 3,000 kilometres a minute or, to put it another way, very roughly 100,000 mph. At that speed you could get to the moon and back in just a little over the time it takes Concorde to touch down on either side of the Atlantic. In human terms the Comet travels nearly 300 miles in the time the world's fastest sprinter takes to cover 100 metres.

Despite its speed it will be hard to detect it actually moving when it is closest to earth in April 1986 because it will still be nearly 40 million miles away.

Add to the possibility of error of this high-speed confusion and of the complicated orbital calculations some other things, such as the tilting of the earth's axis, the phases of the moon, the seasons of the year, possible weather conditions and temperatures, and one realises what a remarkable job scientists and mathematicians have done to tell so accurately in advance the best times to look for Halley's Comet and its position day by day and month by month.

Tips for viewing the comet

These following seven general points are common to observers in both the northern and southern hemispheres.

1. Try to find a place completely dark, get away from city lights and areas where there could be mist, haze or dust. Anything like bright moonlight or pollution of the atmosphere which dilutes the brightness of the Comet will obviously make it difficult if not impossible to see it properly.

2. Consult your daily newspaper which should from the autumn of 1985 to the spring of 1986 give precise directions as regards time and horizon positions of the Comet as well as phases of the moon. In the United Kingdom British Aerospace (in conjunction with Halley's Comet Society) inaugurated in January 1985 the Halley Hotline for anyone with access to a telephone to find out the Comet's position. It is updated frequently. The relevant phone number for your area can be gained by dialling Directory Enquiries.

3. Get yourself a small telescope or powerful binoculars, making sure they are by a reputable maker and have achromatic lenses. This means that it has two lenses close together so that when the light strikes the first, any distortions or colours it transmits are then corrected and re-converted by the second to the original colourless light.

   Although the Comet will be lost from view completely when it is closest to the sun (February 9th, 1986), it can be seen even with the naked eye at certain periods before and after that date (see my chart, pages 36–7) in various countries of the world, except in the more northerly latitudes.

4. When moving from light into darkness give your eyes fifteen to twenty minutes to become accustomed to the change. Then the naked eye will allow one to take in the Comet with one look. The slight movement of the Comet from night to night will be easily detected.

5. The ability of binoculars to magnify the intake of light will bring out better the full impact of the Comet but the field of view is not as wide as that of the naked eye, so you will have to move the binoculars slowly from side to side or slightly up and down to cover its full extent. Among the many sizes and types of binoculars available, 7 × 35 and, better still, 7 × 50 will secure the best viewing.

6. The approaching Comet comes closest to the earth on November 27th, 1985, but then in the weeks that follow the earth draws away from it on its anti-clockwise orbit around the sun, while the Comet continues its clockwise and accelerating thrust towards the sun. This gives viewers in the northern hemisphere a better chance of seeing it than later on when, in early April 1986, the comet on its way back comes even closer to earth as their respective orbital paths once again start to coincide. It is at this time that the most satisfactory observations of the Comet in all its post-perihelion glory are reckoned to be obtainable, particularly by those in the southern hemisphere.

7. Many will wish to have a souvenir of their morning or nightly vigils and comet photography is easy. Any camera whose shutter can be locked open will suffice. Most adjustable cameras have a 'B' shutter setting for this purpose. Some automatic cameras that make exposures of reasonably long duration will also do the trick. But one must place the camera on a strong

and very rigid tripod. Use a cable release to open the shutter with a minimum of vibration. Using fast black and white or colour film on exposures of ten seconds' to ten minutes' duration will photograph the Comet satisfactorily. Because the earth is rotating, the longer exposures will produce significant trailing of the Comet and blurring of background of star images. The lens should be used at the *lowest* f/number possible and one with a focal length of 28–200mm will be adequate for the job.

Northern hemisphere

As a general guide Halley's Comet will be seen through small telescopes during the autumn of 1985.

In late November and December the Comet will be seen, using binoculars, about halfway between the horizon and zenith in the south-west about one to one and a half hours after sunset.

By early January it will be visible to the naked eye and it will brighten rapidly and develop a tail as it gets closer to the sun. During these few weeks before it gets lost in the solar glare at the end of the month (and most if not all of February), it will be seen lower in the sky and more towards due west each night at the end of twilight.

At the beginning of March, the comet will reappear in the *morning* sky about due east and low on the horizon. One will have to get up early, about one to one and a half hours before the dawn light filters in. But this time the tail of the Comet will be bigger.

During March the Comet climbs higher in the morning sky and moves further to the south. At the same time it keeps on getting brighter.

In late March and early April 1986, the Comet is the closest to earth it can ever be until it returns in AD 2061. It will now be at its brightest and showing its greatest length of tail. But it will be seen only low in the south-east and moving further towards the south daily. This will cause distortion by the earth's atmosphere. (This is why people in the southern hemisphere will get the best viewing.) During the last two weeks of April 1986, the Comet might just be seen with the naked eye in the south-east after *evening* twilight. It will start to rise higher in the sky daily but the tail will be decreasing all the time and from May onwards it can only be traced with the aid of binoculars and telescopes as it gradually returns to the deep freeze of the outer solar system.

Southern hemisphere

Here observers will not see the Comet so well early in its appearance as northern observers, but in 1986 the best overall viewing will occur south of the equator. Nevertheless Halley's Comet will be visible through small telescopes during the autumn of 1985. Dr Donald K. Yeomans has indicated that in December the

Comet will be visible with binoculars low on the horizon north of west at the end of evening twilight (about one and a half hours after sunset). He continues:

> In early January observers may see that the Comet has brightened to naked eye visibility in the west. Unfortunately, its motion carries it quickly into the solar glare. The Comet is then invisible for more than a month.
>
> During the last half of February the Comet reappears in the *morning* sky slightly south of due east and low on the horizon before the onset of morning twilight (one to one and a half hours before sunrise). The Comet's motion carries it higher in the sky daily and its tail is soon observed to be better developed than in January. Through the second week of March the Comet fades while the tail continues its growth. The Comet is halfway to the zenith and rising higher daily when it begins brightening after mid-month with its approach to Earth.
>
> In late March and *early April the Comet is closest to earth*. The Comet is now brightest and shows its great extent for this appearance. It is observed about three-quarters of the way to the zenith and is moving rapidly from east-south-east through south to south-west.
>
> By mid-April the Comet has begun to fade and is visible a quarter of the way to the zenith in *both* the morning sky in the south-west and in the evening sky in the south-east. This situation does not last long and a week later Halley's Comet is much better seen high in the *evening* sky.
>
> The Comet has left the morning sky by the end of April and is at the limit of naked eye visibility. For the next several months the Comet can be seen in binoculars or a small telescope as it returns to the deep freeze of the outer solar system.

Finally I thought it would be interesting to provide a set of tables which will enable you to know the Comet's distances in the month of your birth or in the context of any other month for the whole of 1985 and 1986. The tables below show the monthly average distances of the Comet from the sun, from the earth and its speed. The distances are in *millions* of miles and the speed in *mph*.

		1985			1986	
	(Sun)	(Earth)	(mph)	(Sun)	(Earth)	(mph)
January	477,848	402,611	38,479	75,720	126,384	105,484
February	452,548	412,320	39,714	61,403	132,559	115,120
March	425,549	433,348	41,153	87,526	84,303	97,632
April	396,760	450,146	42,831	149,304	58,997	79,050
May	367,875	450,352	44,704	171,743	119,506	67,872
June	337,860	429,335	46,889	211,538	211,591	60,631

	1985 (Sun)	(Earth)	(mph)	1986 (Sun)	(Earth)	(mph)
July	305,479	383,704	49,597	247,436	242,608	55,641
August	271,541	316,942	52,911	282,784	359,351	51,750
September	236,884	236,225	57,004	316,222	404,662	48,649
October	198,821	146,068	62,754	347,109	425,064	46,187
November	158,220	77,662	70,840	376,761	424,208	44,106
December	116,421	81,754	83,676	406,201	407,574	42,261

After February 9th, 1986 (perihelion), Comet and earth get closer for a while but the speed constantly diminishes.

On November 27th, 1985 the incoming Comet will be 58 million miles from earth.

On April 11th, 1986 the outgoing Comet will be only 38 million miles from earth, the closest it will get on its 1985–6 visit.

In the adjoining table I have attempted to offer a guide to the chances of 'naked-eye' viewing, given the right conditions and assuming that the viewer knows where to look. Dr Donald K. Yeomans has cast an expert eye over my forecasts. Although at the time of writing, he says, it is wise to be somewhat pessimistic about 'naked-eye' viewing, the situation may turn out far brighter (in every sense of the word) than we have anticipated.

> It would be prudent [comments Yeomans] to say that during the months of November and December 1985, the comet will be seen only with binoculars and then only if one knows where to look. During the first two weeks of January 1986, the comet will still be a difficult naked eye object and during the last two weeks in January the comet will be too close to the sun for observations. The comet is both intrinsically brighter and with a far more obvious tail after perihelion (February 9th, 1986). In early January observers may be able to observe the comet with the naked eye, but only with difficulty. Late March and early April 1986 are the preferred times for observing the comet with the naked eye.
>
> Before the comet reaches perihelion, the comet's declination is north (positive) so that northern hemisphere observers are favoured. After perihelion, the southern hemisphere observers are favoured.

Dr Yeomans issues a final few words of caution: '*All of these comments should be prefaced with the warning that comets seem to delight in thumbing their noses at earthlings who try to predict their behaviour.*'

Harpur's Guide to Viewing Halley's Comet with the Naked Eye

* Improbable (so have binoculars or small telescopes handy)
** Just possible *** Good **** Very good

	Mid Latitude	Nov 85	Dec 85	Jan 86	Feb 86	Mar 86	Apr 86	May 86
AFRICA (north)	30°N	*	**	***	—	*	**	—
AFRICA (mid)	0°	*	**	**	—	***	***	*
AFRICA (south)	35°S	—	*	*	—	***	****	*
AUSTRALIA	28°S	—	*	*	—	***	****	*
AUSTRIA	50°N	*	*	**	—	*	*	—
BELGIUM	51°N	*	*	**	—	*	*	—
BRAZIL	15°S	*	*	**	—	***	***	*
CANADA	52°N	*	*	**	—	*	*	—
CHILE	30°S	—	*	**	—	***	****	*
CHINA	30°N	**	**	***	—	*	*	—
DENMARK	53°N	*	*	**	—	*	*	—
FINLAND	60°N	—	—	*	—	—	*	—
FRANCE	45°N	*	*	**	—	*	*	—
GERMANY	52°N	*	*	**	—	*	*	—
GREECE	37°N	*	**	***	—	*	*	—
HONG KONG	20°N	*	**	***	—	*	*	—
HUNGARY	45°N	*	*	**	—	*	*	—
INDIA	22°N	*	**	***	—	*	*	—
IRAN	32°N	*	**	***	—	*	*	—
IRELAND	52°N	*	*	**	—	*	*	—
ISRAEL	31°N	**	**	***	—	*	*	—
ITALY	43°N	*	**	***	—	*	*	—
JAPAN	37°N	*	**	***	—	*	*	—
MAURITIUS	23°S	—	*	**	—	***	****	*
NETHER-LANDS	57°N	*	*	**	—	*	*	*

	Mid Latitude	Nov 85	Dec 85	Jan 86	Feb 86	Mar 86	Apr 86	May 86
NEW ZEALAND	43°S	—	*	*	—	***	****	—
NORWAY	63°N	—	*	**	—	—	—	—
PHILIPPINES	14°N	*	**	***	—	*	*	—
PORTUGAL	37°N	*	**	***	—	*	*	—
SAUDI ARABIA	25°N	*	**	***	—	*	*	—
SPAIN	40°N	*	**	***	—	*	*	—
SRI LANKA	7°N	**	**	***	—	*	*	—
SWEDEN	62°N	—	*	**	—	*	*	—
SWITZER-LAND	45°N	*	*	**	—	*	*	—
TURKEY	30°N	**	**	***	—	*	*	—
UNITED KINGDOM	52°N	*	*	**	—	*	*	—
URUGUAY	28°S	—	*	**	—	***	****	*
U S A	35°N	*	*	**	—	*	*	—
U S S R	52°N	*	*	**	—	*	*	—
VENEZUELA	10°N	*	**	***	—	*	*	*
WEST INDIES	17°N	*	**	***	—	*	*	—
YUGOSLAVIA	43°N	*	**	***	—	*	*	—

Note: In February 1986 the Comet is mostly lost in solar glare as the earth is on the wrong side of the sun.

Halley's Comet taken from Lowell Observatory on its last appearance in 1910.

5

What Happened when Halley's Comet Last Appeared, in 1910

I think one can safely say that there will be no diminution on this occasion of the Comet fever which swept the world in 1910. Indeed with the data from space missions already launched and with the Comet being seen literally years ahead of the first sightings achieved in all its previous apparitions, the expectation of the Comet's arrival has ensured an unparalleled level of advance interest.

Even the superstitions, fear, awe and unreasoning apprehension of comets which its previous appearances have revealed within the subconscious of so many of us will not be entirely allayed by the new scientific sedation, the latest mathematical medicine, the therapy of modern technology.

I will be surprised if, just as in centuries past, the soothsayers are not out saying their sooths with the same force as Shakespeare described when Julius Caesar met his death. The humble sandwich-board message 'Prepare to meet thy doom' will no doubt be enhanced by laser displays. An air of unnatural tension will pervade the population provoking extremes of behaviour. There will be suicides. There will be solemn religious cults, and prophets introducing 'second comings' and quoting from the Book of Revelation.

Just as in 1910, on the other hand, there will be much gaiety, characterised by comet parties, star-gazing safaris, comet music, comet products and comet fashions. It will be a period when people react like people who know they are participating in something awe-inspiring and spectacular and which is unlikely to occur again in their lifetime.

We have only to go back to the last appearance of Halley's Comet in 1910 to find the evidence. Indeed there are many who were youngsters then who are still alive and can vividly recall their experiences. The emergence of telegraphy, photography, high-speed distribution of mass-circulation newspapers were crucial factors in enabling comet fever in 1910 to be shared for the first time simultaneously in all continents.

Exhaustive research into British newspapers showed clearly that there was enormous coverage relating to the Comet. This was largely stimulated by exchanges of learned letters between astronomers and scientists and from interested observers. All the newspapers carried daily reports and speculations about the progress of the Comet and from what points in other parts of the world it had been observed.

It was the same in other countries. On May 9th the *New York Times* noted that 'Bermuda observers report comet acting strangely following King Edward's death'.

Beginning with the Saturday edition of May 14th and continuing on through the Sunday edition of May 22nd, the Comet was given top billing. This was the period when the Comet was at the height of its brilliance and activity and the coverage clearly reflected this. Here are some of the headlines during this week:

May 14th: New York City hotels' roofs being used for comet parties; Professor S. A. Mitchell tells of superstitions surrounding comets through the ages in NYC speech.

May 15th (*Sunday edition*): Speculation on probability of earth passing through comet's tail. Article on those still living who remember Halley's Comet visit of seventy-five years ago.

May 16th: European and American astronomers agree earth will not suffer from passing through comet's tail.

May 17th: Earth will pass through comet's 24-million-mile-long tail on May 18th. Hotels to prepare for comet watchers; Boston will sound fire alarm if comet is visible. Editorial comment on fears about comet.

May 18th: Earth to pass through comet tail for six hours; C. B. Harmon invites college deans to join him in viewing comet from balloon.

May 19th: HALLEY'S COMET BRUSHES EARTH WITH ITS TAIL (ran the *New York Times*'s banner headline); 350 American astronomers keep vigil; reactions of fear and prayer repeated; all-night services held in many churches; previous dire prophecies recalled by comet scare.

May 20th: Leading observatories confirm *New York Times*'s discovery of comet tail in east rather than in west.

May 21st: Calculations indicate tail may have passed earth, missing it by 197,000 miles.

As one reads the headlines and the articles in the *New York Times*, or any other newspaper, a feeling of excitement builds, especially during May 1910.

Let us, therefore, go through some of the fascinating things that happened in 1910 in some detail and then retrace the events which coincided on or about all previous appearances of Halley's Comet every seventy-five years or so as far back as 240 BC.

Thanks to Joseph Laufer's Halley's Comet Watch Newsletters one can get a very good idea of the way public anticipation built up from the extracts he quotes from the *New York Times*.

The earliest serious report appeared on August 25th, 1909, approximately eight months prior to perihelion (April 20th, 1910). The headline read: 'Worldwide observatories scan skies for first sighting of Halley's Comet.' The following month five separate stories appeared, including 'Comet reportedly sighted at Heidelberg'. Four more stories appeared in October and four again in November, with a full article and illustration in the Sunday edition on December 5th. On eleven dates in January 1910 the newspaper reported sightings throughout the world and published Comet-inspired editorials on January 25th and 29th. Fear began to mount during February 1910, with three reports about poison cyanogen fallout, and an editorial on this subject on February 11th.

An April 9th report stated: 'Halley's Comet seen by several observatories; not yet visible to naked eye.' The Comet reached perihelion on April 20th (a Wednesday) and the story was headlined: 'Observatories report comet closer; is visible to naked eye in Curaçao.' On April 23rd this item appeared: 'Women and foreigners attribute darkness over Chicago to comet; some become hysterical.' On the same day, a letter to the editor calls attention to the parallel between Mark Twain's life-span and the appearances of Halley's Comet (Twain was born in the year of the Comet's previous visit, 1835, and died on April 21st, 1910, the day after perihelion).

On April 24th an item noted that the appearance of the Comet increased the demand for telescopes in New York City. The latter days of April saw increasing reports. Every issue during May except the 2nd and 4th carried a story on Halley's Comet. It was reported on May 1st that the demand for telescopes to view the Comet almost exhausted the supply in New York City. In a speech reported in the May 5th edition, Professor H. Jacoby said that the earth was in no danger of collision with the Comet. It was at this time that the *New York Times* designated a special reporter for the ongoing story. Miss M. Proctor was to deliver many stories and articles over the next several weeks during the height of the New York viewing of the Comet.

These somewhat banal little snippets only give a hint of the amazing reactions and incidents which erupted all over the world. There was a dual reason in 1910 for millions of ordinary people to get caught up in the emotional maelstrom which Halley's Comet caused. Firstly there was the build-up of its predicted arrival closest to the sun on April 20th aided by mass newspaper publicity and, for the first time, photography; secondly the agony of anticipation was extended to new heights with the revelation that about a month later, May 18th to be exact, the earth would pass through the Comet's tail.

The resulting speculations by experts and others about what would happen included exploding gas, meteoric showers, terrible electrical storms, the earth colliding with the Comet, millions being poisoned by the cyanide mixture in the particles of the tail, a display of light in the upper atmosphere which would literally blind every spectator, and great forebodings of just about every conceivable catastrophe.

Most newspapers tried genuinely to be reassuring but public concern would not be allayed. In Milwaukee for example it was reported that throughout the whole of the 'Fatal Eighteenth' there was a mad rush by thousands to make wills, and the bar, restaurant and hotel business went virtually dead because people were staying at home to be with their loved ones and families . . . just in case! Family reunions on a scale which normally take place only for births, deaths and marriages were evident everywhere.

A number of suicides were attributed to the impending event and one unfortunate woman went mad from religious hallucinations brought about by the belief that the Comet was going to destroy the world. Farmers in one area took down their lightning conductors so as not to attract the Comet. Several girls in a Milwaukee suburb buried their love letters to protect their amorous secrets should they die. Prayer meetings were rife and by Lake Superior people abandoned their homes in case the impact of the Comet's tail caused a tidal wave.

The following headlines and résumé of stories which appeared in newspapers all over the world present a bizarre reflection of human behaviour, with tragedy,

Study the detail of this powerful drawing of 1910 by Arnold Moreaux making the point that with a gigantic tidal wave one could only find refuge on top of a mountain.

comedy and extraordinary coincidence all linked one way or another with the Comet's spell-binding presence.

Extracts from the May 1910 issues of the *Washington Post* provide extraordinary evidence of the human fears and tragedies:

> Suicides have been reported before because of fear at the approach of Halley's Comet, but Bessie Bradley, 25 years old, committed suicide to-day at Hastings-on-Hudson, where she was employed as a maid, because the comet *failed* to appear.
>
> The young woman became greatly worried over the contradictory reports of what the comet would do and what it would not do. When no sign of it showed last night she grew so nervous that she could not sleep, and this morning she went to her room to rest. Other servants who sought to rouse her for luncheon found her dead on her bed, and the gas turned on.
>
> Miss Kate Van Ness, 40, of Carlton Hill, was taken to the Morris Plains Insane Asylum to-day by Constable Harry Dawson, of Hackensack. The unfortunate woman was a victim of nervous collapse following the comet agitation. All the way to Morris Plains she continually said she would follow the comet no matter where it went.
>
> Millie Morris, colored, was one of a crowd which gathered at the bridge over the Rappahannock last night to watch for the comet. She expressed great alarm over the effects of the comet's visit and then fell dead.
>
> W. J. Lord (of Alabama) is in a precarious condition as the result of four alleged attempts to commit suicide. With his mind wrought up over the proximity of Halley's comet, it is stated, and believing that he had sinned against the Holy Ghost, Lord is said to have made an attempt to shoot himself. Unsuccessful in this, he jumped off a roof and fell on his head, knocking out his teeth and sustaining other injuries. He then cut his throat and jumped into a well.
>
> Talladega, Alabama: The appearance of the comet this evening caused intense excitement here. Congregations of several churches left their pews and hundreds of persons stood excited in the square and gazed at the celestial visitor.
>
> Miss Ruth Jordan, daughter of a farmer living 2 miles from here, was called to the door of her home to see the comet and immediately fell dead, physicians assigning heart failure as the cause.
>
> An unknown negro on the depot platform was shown the comet and instantly dropped dead.
>
> In New Jersey James Kline, a negro, formerly a Pullman car porter, is in the Somerset county jail here a raving maniac as the result of waiting in terror for five days for the destruction of the world by Halley's comet. A

policeman was standing on Main Street early this morning, when Kline in scant attire ran past him shouting that he was being pursued by his mother-in-law and the tail of Halley's comet.

When the officer shouted, the fleeing negro stopped suddenly and began to pray. Kline is sober and industrious. About a week ago he went through the negro colony warning his colored brethren to prepare for the end of the world.

Pittsburg: Reports on church attendance made to-day by eighteen ministers of Pittsburg show an increase of 30 per cent in the past two weeks.

Many persons who haven't been in church for years have been in their pews, and until after a general discussion the cause for the increase was unfathomable. After much discussion the majority of the ministers declared that the Rev. F. A. Wright, pastor of the Fourth Christian Church, is directly responsible.

Three weeks ago Dr Wright declared that the Halley comet was a forecast of the second coming of Christ. This prediction is said to have been taken seriously.

San Bernardino, California: While brooding over possible ill effects of the comet's visit, Paul Hammerton, a sheepman and prospector, became insane and crucified himself, according to mining men, who arrived here with him yesterday. Hammerton was found where he had nailed his feet and one hand to a rude cross, which he had erected.

Although he was suffering intense agony, Hammerton pleaded with his rescuers to let him remain on his cross.

Since the visit of an earlier comet in 1910 Hammerton has been much alarmed, and when he learned that the earth was scheduled to pass through the tail of Halley's comet his mind gave way. He believes that the end of the world is at hand.

Pitcher, N.Y., May 14: Friday the 13th was not such an unlucky day on the farm of Amos Rhoades. This morning one of his prize Dorsetshire cows gave birth to four well-formed calves.

Two of them have star-shaped markings on their foreheads, which, the dairy-maid says, are due to Halley's comet.

The *Chicago Tribune* of May 17th, 1910 carried a report from Paris illustrating how the Comet got blamed for everything:

Paris, May 16: The Bicetre Hospital here was the scene to-day of a terrific explosion. The walls and staircases were wrecked and the inmates, who are all aged persons, became so imbued with the idea that the comet had struck the earth that they made a rush for the gates. It was with extreme difficulty that they were driven back.

An investigation showed that the explosion was caused by an infirmary keeper, who was experimenting with nitroglycerin and who was killed. There

Left: A dramatic interpretation by the artist Koek-Koek in *Pearson's Magazine*, 1910 on the theme of the world being doomed if a large comet came too close. The choice of St. Paul's Cathedral being destroyed was ironic as St. Paul's was rebuilt after the fire of London (1666) to designs by Christopher Wren who was a contemporary and friend of Edmond Halley.

Below: A German cartoon of 1910 anticipating a possible collision with the Comet as the tearful earth indicates. By August Roeseler from *Fliegende Blätter*.

Right: The Comet discarding its comb and hairpins highlights the celestial publicity it has received. The drawing from *Le Rire,* May 1910, pokes fun at the use of the Comet in advertisements.

Below: Joy-riders of space, an animated scene with a speeding automobile (a rare phenomenon, too, in 1910) possibly representing the Comet. Cartoon from *Puck's Monthly,* May 1910.

Continental postcards and drawings of 1910 show the popular concern about the Comet signalling the end of the world.

were enough explosives in his room to blow up Paris, but they did not go off when the nitroglycerin let go. The man was haunted with the idea that he was an inventor.

Further bizarre stories like these were carried by the *New York American*.

Michael Sweeney, of No. 204, East Twenty-First Street, says he is through with the comet for ever.

Last night he was returning home with his pay envelope containing $27.50 in his pocket. While crossing Second Avenue his attention was attracted by a crowd of men looking out of the windows of the Anawanda Club, Charley Murphy's organisation in the Gas House district, and heard them calling to other men on the street to watch out for Halley's comet.

'I looked up and down the street,' Sweeney later told Captain Surfeind, of the East Twenty-Second Street police station, 'for I thought something was coming. Then I saw a lot of children looking at charts and I knew it was nothing that would interest a stone cutter. I hurried home and discovered that my pocket had been cut and my money taken.' 'Sure, and it was Halley's comet that done it, for I have lived in this Gas House district fifty years and never before lost a cent,' was his final comment, as he departed with the two detectives in search of a clue to both the pay envelope and the comet.

Woodbury, N.J., May 7: So that any one who wishes to see Halley's comet need not stay up all night, Mayor Ladd has instructed the police to be on the lookout for the sky visitor, and to notify by telephone all those who want to be awakened. A dozen families were called up at 3 o'clock this morning, when the comet was visible. 'Get up and see the comet' was the call.

> Vienna, April 23: An amusing account of the way in which the inhabitants of a small Hungarian village prepared for the end of the world has reached Vienna. In a village in the Theiss Valley, the inhabitants have been expecting the end of the world for some weeks, believing that on the appearance of Halley's comet the whole globe will be smashed to atoms.
>
> Some days ago a large fire broke out toward midnight in a neighbouring village. The watchman, seeing the skies lighted up, walked through the streets blowing his horn to rouse the inhabitants and shouting 'The last day has come!' The people rushed half-clothed from their abodes to die in the open. Men trembled, women screamed, and the children cried.
>
> What followed was a curious satire on the actions and thoughts attributed to the dying by writers of poetry and fiction. The simple people considered first that all the provisions in the village should be consumed. A large fire was lighted in the square in front of the church, and there food and drink were brought out of the houses. Everyone joined in a hurried orgy, while hasty prayers were made between bites for the salvation of their souls.

> A number of Towaco, New Jersey, farmers are bewailing the loss of chickens stolen early yesterday, while they and their families were on top of Waukhaw Mountain, waiting for Halley's comet to appear.
>
> Nehemiah Doolittle yesterday said two well-dressed young men, self-styled scientists, drove through Towaco Thursday afternoon, spreading the news that the comet would be nearest the earth at 3 am Friday, very bright, and the whole of its tail would be visible.
>
> They offered ten, five and two-and-a-half dollar gold pieces for the best description of the comet.
>
> By 2 am yesterday everybody in Towaco was on the mountain top. Not a glimpse of the comet did they get. In their absence practically every chicken coop in Towaco had been raided.

The universal impact of the heavenly phenomenon was confirmed in London's *Daily Mail* by the simple opening sentence of a story in May. 'In every part of the world Halley's Comet is at present a subject of unique interest.' Under the headline '*Mademoiselle Halley*', its Paris correspondent reported:

> 'Comet suppers' are to be in vogue in Paris on Wednesday night, when Halley's Comet is expected to show traces of its presence over the boulevards in its passage across the sun's disc.
>
> For days past, hawkers have sold hundreds of picture postcards illustrating the end of the world on the arrival of 'Mademoiselle Halley', as the Comet has been nicknamed. Brooches and pins showing the Comet with its tail, and special newspapers parodying the imaginary catastrophes which the astral visitor will bring with its tail, have been widely sold. The newspapers publish long details of observations of the Comet made at French and foreign observatories.

What Edmond Halley would have thought about his Comet being dubbed 'Mademoiselle Halley' is probably predictable. He was a well-travelled man who knew the French, and would have appreciated that the characteristics of a comet with its long fine 'hair' and an ability to attract attention with its glowing magnetism must make it female. However the fickleness of woman was suggested by a subsequent report from the same Paris correspondent under the headline '*Disappointment in Paris*'.

> Paris is vexed with the Comet. Instead of appearing radiant and beautiful in the skies last night 'Mlle Halley', as the Parisians firmly believe, sent a torrential downpour accompanied by thunder, obscuring the heavens, so that nothing whatever was seen of the Comet. Balloon ascents were abandoned, and those energetic Parisians who spent the night on the summit of the Eiffel Tower, or clambered up the steep steps to the Sacred Heart Church, on the top of Montmartre, had their pains for nothing.

Newspapers reported a '*Gay Vigil in Madrid*' with

> an incessant stream of people passing through the streets proceeding to elevated points to view the Comet, which however, will probably not be visible as the sky is overcast.
>
> With the tone of gaiety mingles a comic note. Passing in the throng may be seen, with the inevitable gaping following, quite a number of astrologers,

> with long-pointed headgear and black tunics, waving their dim torches. Armed with long measures, they are doubtless going to take the scale of the mysterious astral body.
>
> Reports from all parts of Spain announce that the people are celebrating with great joy the transit of the Comet.

In Italy 'Animated Scenes' were reported from Rome when cafés and restaurants remained open twenty-four hours a day just as they did for New Year's Eve. News stories from all over the world indicated, just like this one, that for every thousand who anticipated the Comet with fear and foreboding there were many thousands more who looked forward to it as an excuse to have one hell of a party.

There were social events galore. People organised Comet Clubs, Comet dances, Comet-watching parties and excursions; there was even a Comet cocktail. One can imagine that with ingredients such as a good dash of vermouth, a stiff tot of applejack, poured over cracked ice, this went down a treat. An article in the *Daily Mail* (from its New York correspondent) reflected the mood in the United States:

These two romantic vignettes (left and facing page) remind one that in 1910, when the permissive society had pushed its frontier to the revelation of a shapely ankle, it must have been a great relief to steal a caress from a loved one under the cover of alleged comet-watching.

America is reading, talking and joking about little else than Halley's Comet. As for Comet parties, their name is legion. The hotel roofs are crowded nightly, and festivity reigns till the dawn of the morrow. Generally speaking, the mornings have been so cloudy that the celestial visitor has seldom been seen. Scores of men are giving breakfast parties at the leading hotels in overcoats, 'on the tiles' with appropriate orchestral accompaniment. In one or two of the more elaborate affairs, small silver telescopes were given to each guest as souvenirs.

Every morning, before sunrise, personally conducted groups of school-children may be seen star-gazing in Central Park. The Mayor of Middle-town, Connecticut, has issued a municipal edict, directing that a general fire alarm shall be rung at 2.30 o'clock every cloudless morning.

This cartoon by Leonard Raven-Hill in *Punch*, 1910, enshrines something of the mutual mystique of the two flying machines, the celestial one seen for over two thousand years and the other barely off the drawing board of the Wright Brothers.

> One New York paper has asked a variety of public persons what they would do if it were absolutely certain that the Comet was going to destroy the earth in three days. Mr de Wolf Hopper says that he would charter the swiftest aeroplane afloat and watch the coming cataclysm from the clouds.
>
> Miss Nora Bayes, the actress who is now singing America's most popular song, 'Has anybody here seen Kelly?' says: 'I should stop doing everything else and stick close to my husband. When the time came I'd put my arms tight around his neck, and then I wouldn't care if the Comet did kill us, and neither would he.'

I made a detailed check of all the stories appearing in *The Enquirer* (Cincinnati) in April and May 1910 when the newspaper published a bulletin giving details of the Comet and when and where it could be observed. The news stories carried were very indicative of the sensational and the trivia on which Comet mania was sustained.

A most striking example is encapsulated in the following report, especially foreboding in tone when one reads the last paragraph containing the grim prophecy of World War I.

> TAIL OF THE COMET. *Turned Red the Night King Edward Died – Scared Bermuda Negroes.* Special dispatch to the Enquirer: Steamship *Bermudian*, at Sea, via Wireless, May 8 – On Friday night, that of the death of King Edward, and, of course, the accession of King George, a remarkable phenomenon was observed at Bermuda and recorded by an experimental scientist.
>
> Halley's comet, which first became visible about 2 o'clock in the morning, later displayed a most decided red tinge on the tail. At 12.30 the fort near the city of Hamilton began firing the salute of 101 guns to the new King at intervals of two minutes. Just as the last gun was fired, at exactly 3.52 a.m., at the report of the last gun, the comet's tail flared a decided red end. The head, now distinctly visible, became a ball of red fire.
>
> This phenomenon lasted only five minutes, but was observed by many negroes working on the docks loading the steamer *Bermudian*, who fell on their knees and prayed, refusing to work.
>
> Does this signify war in George's reign? Shall some great calamity befall the earth? Sailing on both the 11th and the 18th of this month, large parties of scientists will come to observe the comet, Bermuda being the most favorable American spot.

Then we had a story headed:

> COMET PILLS. *Being Sold to Haitian Negroes by Aged 'Voodoo Doctor'.* New York, May 16: Whatever the comet may do, the negroes of Port au Prince, Haiti, are prepared. They know they are safe because they are well stocked with comet pills.
>
> Comet pills are new to the pharmacopoeia. Word of their appearance arrived by the Hamburg-American Liner *Allegheny*, in from Port au Prince, to-day. Her officers said that all the negro stevedores there, all the farmers

> roundabout, the servants, laborers, merchants, beggarmen and thieves are rushing to the hut of a shrewd old voodoo doctor just outside the city, who is selling comet pills as fast as he can make them.
>
> The prescription is one pill for every hour up to the time the comet begins to recede from the earth, but many of the patients are making safety doubly sure by taking one pill every half hour. The comet doctor guards the formula closely, and is growing rich.

This item has inspired the manufacture of comet pills once again – but this time in a more laudable cause, in order to raise funds for a New Jersey museum.

Perhaps one of the most lurid and horrific tales was that of a beautiful Oklahoma girl who was rescued in the nick of time by a Sheriff's posse from being offered as a blood sacrifice in atonement for the sins of the world. It was stated that a Miss Jane Warfield (aged sixteen) was saved from slaughter by a band of religious fanatics called the 'Select Followers'. Their leader, Henry Heinman, said he had received a revelation from God that the world was to end on May 18th, 1910, and the 'heavens would be rolled up like a scroll following the contact with the tail of the Comet, and to save the world a blood sacrifice of a young maiden was needed'. She was clad in spotless white with a wreath of white flowers round her head. Her hands were bound and Heinman was standing in front of her with a long sharp hunting knife raised high for the first strike when the posse arrived only just in time.

This story is now thoroughly discredited because, although it was carried by many newspapers all over the USA with varying versions of luridness, the one significant omission was the local newspaper in whose area the dramatic incident was alleged to have occurred. It never mentioned it.

The brief headings of other stories underline the range of reactions and human emotions as the Comet approached.

'The Comet is here, shrieked Dayton woman when auto crashed into her hens.'

'New Orleans. Many rush in places of shelter crying out for mercy and kneeling in prayer.'

'Powder mills closed for fear gases might cause explosion.'

'Churches packed with excited foreigners in Chicago – others lock selves in cellars.'

'Atlanta negroes refuse to work fearing Comet will destroy the earth.'

'Comet balking Cupid – Chicago clerk explained decrease in licences.' (He said, 'They think what's the use of getting married only to be wiped out of existence.')

'Comet blamed for mental derangement of Ohio man who watched for it.'

'Glass balls fell in Kentucky during rainstorm and Comet blamed for phenomenon.'

But perhaps the most human and amusing story of all is one of unashamed 'voyeurism'. To understand the story one has to recall that when an amorous couple were flirting and embracing some fifty years or more ago they were what was then called 'spooning'. So this story had a headline: 'Comet Forgotten – when Villagers saw "Spoons" on Hillside Through Telescope.' The dateline was Buffalo, NY, May 13th, 1910 and went on:

A classic Heath Robinson with the gentleman adding the final touch of magnification by dangling his glasses at the top.

STAYED UP TO SEE IT.

NOCTURNAL WANDERER.—Orísher, 'blige me! Whish one er thosh ish Halleysh Comet?

Burglar (with sudden enthusiasm for astronomy). "'SCUSE ME, GUV'NER, CAN YOU TELL ME WHERE I CAN GET A VIEW OF THIS 'ERE COMET?"

> The Comet has struck the little neighbourhood village of Gowanda. [The story is now transferred from the bottom of the front page, not less, to an inside page with the heading 'FLIRTING'.] Not a comet, however, for which some of the aggressive residents of the villages were looking early this morning through a big telescope bought for public use, but the effects upon the village bid fair to be none the less revolutionary than if the heavenly comet had landed.
>
> A group of prominent villagers was peering through the telescope this morning, when the apparatus suddenly focused itself upon a hill. Immediately there appeared on the lens the figures of two people, said to be well known, in fond embrace.
>
> Immediately the elderly men behind the telescope held a conference and decided to forsake the search for the heavenly body and devote the remainder of the morning looking for terrestrial bodies. Much to the amazement of all it is said that exactly six other similar scenes were brought upon the lens.
>
> Report from Gowanda to-night is that the village is greatly perturbed by the unexpected discoveries of the amateur men of science. It is even said that divorce suits in the near future are not regarded as improbable.

The world's preoccupation with the Comet in 1910 was not only reflected in the newspaper reports and frenetic social activities, but also found expression in countless cartoons, postcards, verses and varied illustrations, examples of which are given in these pages.

In fact there was not a facet of any creative medium neglected in the enthusiasm of people all over the world to ventilate their inspiration. This of course included music. Many compositions were just named 'The Comet' and took the form of marches, two-steps, waltzes, Comet 'Rags' (one such 'Rag' had a subtitle which styled it, flirtatiously we think, as 'a continuous "come-on"'!) and one was called simply 'Comet Schottische', followed by the explanation, 'Latest Society Dance'. Ah! what an era of romance and nostalgia came to an end when the death of King Edward VII of England and the arrival of Halley's Comet coincided.

Alas, none of the compositions seems to have got into the charts at any time, at least not to survive the passage of the years as did 'Has anybody here seen Kelly?'.

Perhaps the search for Kelly was more inspiring than Comet watching because apart from a lack of tuneful originality very few Comet pieces had lyrics, and those that had were in the vein of the following extract, which did little to promote a rush of blood:

> Halley's Comet, we heard about you,
> Now we can see you in the sky so blue,
> People say that on the eighteenth day of May
> We did sail through your long tail.
> But it was so very pale, and so very frail,
> That it didn't do us any harm to sail
> through your long tail.

Any review of the threads of the massive tapestry of events which the 1910 app[illegible]
provided must include the remarkable claim of Mr E. W. Ryall of Benfleet, Essex, in England. He said in a letter to me that from his memories of a previous life which he published in a book, *Second Time Round* (published by Neville Spearman, London), referring to his existence in the seventeenth century,

> I am convinced that I have witnessed Halley's Comet on two of its appearances in 1682 and 1910. I feel I should make it clear to you that I cannot claim any better insight than yourself into the periodic visitations of this phenomenon, as to its path through the heavens, its predictable reappearances within sight from this planet, and its effect (if any) upon the affairs of this world.

While there are quite a few people who were of a tender age in 1910 and who will probably see the Comet again in 1986 there can be little doubt that Mr Ryall's sighting of the same apparition some 228 years apart must be unique, even allowing for those who believe in reincarnation.

In the next chapter we take a brief look at the commercial aspects of the Comet in 1910 (some of the products are still going strong) and after that, although we may still be reeling from the impact of the 1910 experience, it seems logical to go back over the other visits to 240 BC and find out how the world was then.

6

How the 1910 Comet Went 'Commercial'

It did not take long for entrepreneurs, ranging from prestigious manufacturers of fine products to general stores and the humble street hawkers, to find ways of selling extra goods and services by cashing in on the Comet fever.

The range and ingenuity of inducements by way of souvenirs were remarkable. On the one hand there were the makers of the finest champagne and quality products using a 'Comet' theme, or retail store owners advertising 'Seats for Comet Gazers' and on the other hand a range of inexpensive souvenirs from postcards to 'Comet pills' (as already described) to tempt the public.

The whole spectrum of public concern and commercial exploitation has been neatly summarised and illustrated by Ruth Freitag in the *Quarterly Journal* of the Library of Congress, Washington, DC, in 1983.

> Halley's Comet in the spring of 1910 was a media event. Comet stories vied for newspaper space with reports on the travels of ex-President Teddy Roosevelt, the floods in Paris, the eruption of Mount Etna, the devastating earthquake in Costa Rica, and the deaths of Mark Twain and King Edward VII. Indeed, the comet was often blamed for these sensational occurrences.
>
> Some astronomers spoke of the possible extinction of life when the planet encountered the comet's tail on the night of May 18, and speculated on the terrible destruction that the comet would inflict should it actually strike the earth. A minister in Pittsburgh filled the city's churches when he announced that the comet portended the second coming of Christ. Although many authoritative voices were raised to counter these frightening statements, the doomsayers were widely believed, and stories of persons driven to madness or suicide were common. Others prepared for the worst by retreating to their cellars with a bottle of oxygen, or, in the great tradition of 'eat, drink, and be merry', partied all night. The general agitation was exploited by swindlers, who sold comet insurance on Capitol Hill and comet pills in Port-au-Prince. Those who are inclined to laugh at the absurdities of yesteryear may be somewhat chastened by reflecting upon the recent 'Jupiter effect' scare. [This refers to the prophecies of doom associated with a certain alignment of the planets in early 1982.]
>
> The comet inspired reams of newspaper verse, comic cartoons and postcards, jokes, short musical compositions and revue sketches. Even Gus Edwards produced a song, 'The Comet and the Earth', for the Ziegfeld Follies of 1910. A cartoon in the *Portland Oregonian* suggested that the

and night every day of the year all over the world and has been since 1789, in the homes of discerning people.

Little has changed. Even in 1910 products of all kinds were linked to the Comet in a bid to increase sales.

nation's professional funnymen should offer the comet a vote of thanks. Today Halley's Comet T-shirts are being sold in anticipation of the coming reappearance, but in 1910 comet designs decorated vests, neckties, handkerchiefs and socks.

The manifestation of interest in the 1910 visit of Halley's Comet that is perhaps least familiar today was the application of the comet motif to advertisements. A sweeping arc of tail, growing out of a bright circular or star-shaped head that usually featured the product, formed an eye-catching design that was used to good effect by many commercial artists.

7

The World's Major Events Coinciding with the Appearances of Halley's Comet since 240 BC

It is intriguing to think that such men as Nero, St Paul, Attila the Hun and Mohammed would have all seen Halley's Comet. It is the Comet which provides a unique link between certain events and people, throughout history. The Comet that hung over Jerusalem 'like a sword' in AD 66, that blazed forth as a bad omen to King Harold in 1066 and which arrived just after the death of King Edward VII in 1910, will be seen by millions of people all over the world in 1986.

It is tempting to try to establish a link between the return of Halley's Comet with a specific historical event such as an earthquake or the death of a great man. While, as you will see, there is plenty of evidence that the arrival of our galactic gipsy has coincided with dramatic events, one appreciates that the turbulence and troubles of the world are so numerous (has there ever been a decade when a war was not taking place somewhere?) that the Comet would be hard put to it to find a year when some trauma was not taking place.

So in this chapter a survey has been made of what happened not just in the years Halley's Comet actually appeared but also the years either side of them. The aim is to build up a fuller historical picture of the state of the world than would otherwise be conveyed by the record of just the year of the Comet's return. So, in AD 66, the Comet was seen over Jerusalem; but it is also interesting to note that a few years before it returned Queen Boadicea was fighting the Romans and one year after it returned St Paul was executed. Likewise, when the Comet returned in 1378, it came two years after the death of the Black Prince and three years before the peasants' uprising led by Wat Tyler. To see how the world changes every seventy-five to seventy-nine years (the parameters of the Comet's periodic visits) is an exciting and intriguing way of looking at history.

240 BC Chinese astronomers carefully record appearance of a spectacular Comet. It is later identified as the same one whose return Halley predicted.

The same year Rome annexes Sicily and puts a heavy burden of taxation upon her inhabitants.

The Cyrenian, Eratosthenes, becomes librarian at Alexandria. He is a versatile scientist. He mapped out the River Nile's course and made a relatively accurate estimate of the earth's circumference.

ng-ti is thought to have begun plans for building the Great Wall of designed to keep out would-be invaders.

Andronicus, once a Greek slave and now a Roman playwright and , puts on his first tragedy in Rome.

the adjacent years, the last year of the First Punic war between Rome d Carthage, when the Romans win a decisive naval battle and force the Carthaginians into submission, takes place in 241 BC. The same year the peace terms are agreed upon by Rome and Carthage: Carthage must give up its intention to occupy Sicily and must not encroach upon Roman territorial waters. In return, the Romans let the Carthaginian troops go home unharmed.

In 239 BC there is civil war in Carthage after mercenary soldiers revolt against the Carthaginian rulers, because they feel they have been badly treated. The revolt is finally quashed later by Hamilcar, the father of Hannibal who was so noted for using elephants to cross the Alps and carry the war against the Romans into Italy itself.

164 BC No detailed mention of this apparition appears in the Chinese records but it is known that a Comet, assumed generally to be Halley's, had its perihelion in the autumn of that year, recently confirmed by Babylonian tablets (see pages 64–5).

Judas Maccabeus*, the great Jewish leader, has refused to stop fighting in return for religious freedom for his people and continues to persecute those who had put up Greek idols to worship in place of the outlawed Judaism.

Antiochus V is murdered by Demetrius who succeeds him as king. Demetrius, who was once a hostage of the Romans, will reign for twelve years as Seleucid King of Syria.

It is probable that the Roman Cato the Elder was writing his book on agriculture at this time. He is now seventy years old and will repeatedly urge the Romans to destroy Carthage – an event that will happen in 146 BC.

Terence, the great Roman comic playwright, is twenty-four years old and has already written three plays. He is now writing *The Eunuch*.

Hipparchus, the Greek inventor of trigonometry, is born about this time.

87 BC The Romans take Athens in the spring. Aristion, the Tyrant of Athens, is executed.

Battle of Chaeronea: the Romans under Sulla decisively beat Archelaus and suffer a remarkably low loss of life.

In Rome, Cinna, a Consul, is forced to leave the city after a riot breaks out

*Three years later in 161 BC Judas Maccabeus dies in battle. He is succeeded as leader of the Jewish resistance by his youngest brother, Jonathan. Jonathan will continue to fight until he dies in 143 BC.

as a result of his own troublemaking. He joins forces with the outlawed Marius. They return to Rome and set themselves up as Consuls. For five days they organise a slaughter of the patricians. Several days later, Marius dies of pleurisy.

Julius Caesar, who is only fifteen years old, is now *flamen Dialis*, priest of Jupiter. He will be stabbed to death forty-two years later.

Sallust, the Roman historian and supporter of Julius Caesar, is born.

In the two preceding years Mithridates, the King of Pontus, takes advantage of Rome's domestic troubles and makes incursions into Roman territories in Asia, taking the town of Pergamum. Some 80,000 Roman merchants and officials are slaughtered.

In Rome, Sulla, the newly elected Consul, is entrusted with the task of doing battle with Mithridates. Sulla's rival, Marius, is exiled from Rome after fighting breaks out between supporters of the two men.

Archelaus, a general of Mithridates, heads towards the west with an army. Athens decides to side with Mithridates against the Romans.

Sulla arrives at Athens and besieges the city.

In the years that followed from about 85-83 BC, there is the Battle of Orchomenus: Sulla again defeats Mithridates. Peace terms are agreed upon by Sulla and Mithridates. Sulla then begins the journey back to Rome.

In Rome, Cinna is ruling. He gets the Senate, which is made up of his own followers, to outlaw Sulla. Eventually however, Cinna meets his death at the hands of his own troops.

Sulla returns to Italy via Brindisi with 40,000 men. He suppresses his opponents in a bloody civil war.

Julius Caesar marries Cornelia, the daughter of Cinna.

12 BC (Sometimes given as 11 BC) Dion Cassius, the Roman historian, reports that a Comet hangs over Rome. The Comet is observed by both Chinese and Roman astronomers. In this and in the immediately adjacent years we find in chronological order that Rome has become the centre of historical events, where Augustus is in power.

Agrippa, a Roman general who is loyal to Augustus, sets out to stop the Parthians from increasing their influence around the Bosphorus.

Tiberius, who will become the second Roman Emperor, and Drusus, his brother, become Praetor and Quaestor respectively, important posts in the Roman Senate.

Agrippa arrives at Jerusalem having successfully dealt with the Parthians. He visits the Temple and pays his respects to Herod the Great who is keen to keep in with the Romans. Herod the Great was granted the title of King of Judaea by the Romans through the efforts of Mark Anthony. It is Herod

EXCLUSIVE

Secrets of the tablets decoded 164 BC Comet sighting confirmed

At first glance this could be taken as little more than a boring picture of four anonymous individuals with one holding two broken pieces of stone in his hands. Do not be deceived. Behind its apparent banality there hides a sensational story and I was lucky enough to be one of the very first and few to be let into the secret which enabled me to organise this historic picture.

It is the first portrait of the four wise men taken together whose combined genius (and genius has been described correctly in this case as an 'infinite capacity for taking pains') offered the world in 1985 the vital evidence confirming the Comet's appearance 2150 years ago.

For this I am indebted to Christopher Walker, Assistant Keeper of the Department of Western Asiastic Discoveries (the one holding the stones) who subsequently introduced me to (from the left) Kevin Yau and Richard Stephenson of Durham University, and Hermann Hunger of the University of Vienna.

We have to thank Chinese astronomers for records of sightings going back to 240 BC and even before that. For over two thousand years the only gaps that existed were those positively identifying the Comet's appearance in 164 BC and supplementing the scanty record of that of 87 BC. But among the thousand and more Babylonian tablets dealing with astronomy which were first catalogued in the British Museum by Theophilus Goldridge Pinches (1856–1934), and whose piecing together subsequently has been aptly described by Mr Walker as the 'biggest jigsaw in the world,' there are two from which the

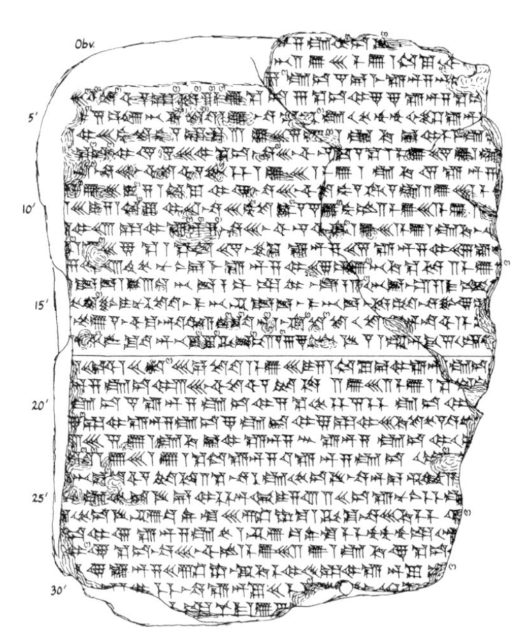

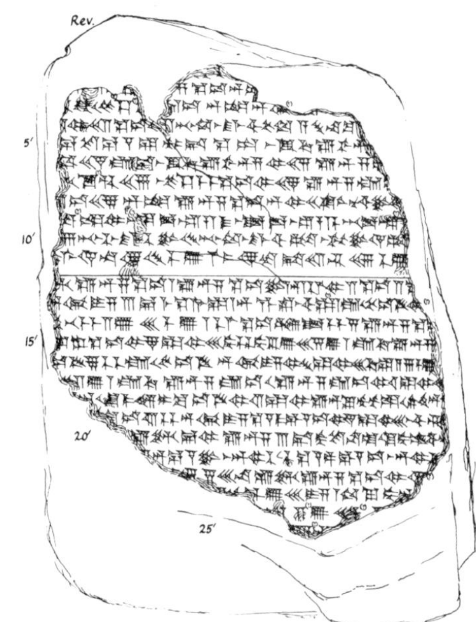

first factual confirmation of the 164 BC visitation has been found. From another comes additional evidence to supplement the Chinese reference to the 87 BC Comet (see a further photograph in the colour section).

The actual writing on the ancient tablet in Mr Walker's left-hand is reproduced to show the amazing symmetry and meticulous spacing of the cuneiform language of the Babylonians at that time. Translated it reads:

> 'The Comet which before had been seen in the east in the path of Anu in the area of Pleiades and Taneus, to the west [here a piece is missing] . . . and passed along in the path of Ea . . .'

Dr Stephenson assisted by Mr Yau has checked the tablet dating for the reference (September/October) 164 BC with experts including Dr Donald Yeomans (to whom I have referred earlier) and there is no doubt in anyone's mind that this is a definitive confirmation of Halley's Comet.

But that is only half the story.

In the early 1950s an American scholar, Professor Sachs, examined the British Museum collection, in particular those tablets called the 'astronomical diaries', which not only recorded astronomical events, but also gave for each month the weather, market prices and local politics. When Professor Sachs died in 1982 his pioneering work was passed on to Professor Hunger, one of the few available experts on Babylonian Astronomy, to be prepared for publication. Then fate stepped in.

In 1984 Dr Stephenson, familiar with Babylonian lunar records and fluent in Chinese was approached about the translation of Far Eastern texts relating to Halley's Comet. Having finished that assignment, but with his curiosity aroused, he was inspired to ask Professor Hunger if similar Babylonian observations might even have come to light. Professor Hunger searched through Professor Sachs' papers and behold, history was revealed. To everyone's delight he found references in three tablets. Two were datable to 164 BC and the other to 87 BC.

To think that the vital link with the definitive record of the 164 BC appearance written over 2,000 years ago should be found in the British Museum within a mile or so of where Edmond Halley was born, seems to stretch the arm of coincidence out of its socket.

who will order the massacre of the Innocents at the time of the birth of Jesus.

The main point of interest about this return of the Comet is that some efforts have been made to justify it as the 'Star in the East' which is only described in the Gospel of St Matthew (Chapter 2) and which inspired the Wise Men to ask Herod, 'Where is he that is born King of the Jews? For we have seen his star in the east and have come to worship him.' The painting of the Nativity by Giotto (see 1301) no doubt helped to fortify the myth when he used Halley's Comet to give a realistic impression of the 'star'. However no matching of the dates of 12 or 11 BC and that of the birth of Christ can be contrived to give a semblance of credibility to any coincidence.

For centuries people have indulged in wishful thinking that it might have been Halley's Comet the Magi (Wise Men) followed. What then are the origins of the myth? Let us take the word 'magi'. It comes from the Persian meaning people with 'magi-cal' powers to read the stars: wise men, astrologers, who had the gift of prophecy. From it our word magician is derived.

Only in the Gospel according to St Matthew is a star mentioned as leading the Magi to the birthplace of the baby Christ. If this was a comet of such significance it is curious that the other parts of the New Testament omit it. The most likely theory is that Halley's Comet of AD 66, which was described by Josephus so dramatically as a 'sword' suspended over the Holy City before it fell to the Roman invaders, coincided with the time when St Matthew's account was being written, and therefore inspired some poetic licence in the interpretation of the Nativity.

Although comets were regarded as evil omens, portents of plagues and the death of kings, there were some from the earliest of times who saw them as signs of great beneficial changes in the order of things and with the *birth* of kings. Origen, a theologian of the third century, wrote in his treatise 'Against Celcius': 'The star was seen in the east . . . we consider [it] to have been a new star . . . such as comets or those meteors [at that time both were regarded as the same] which resemble beams of wood, or beards or wine jars.' He then went on to explain that when something good was going to happen comets could make an appearance, and what better time than at the birth of Christ when the world would be changed. This he substantiated with an Old Testament quotation (Numbers 24:17) 'There shall arise a star out of Jacob and a sceptre shall rise up out of Israel.' This is given startling relevance in the *New English Bible*, which translates it as 'a star shall come forth out of Jacob, a *comet* shall arise from Israel.'*

The confusion is inevitably compounded by Giotto di Bondone who saw Halley's Comet in 1301, when it had a particularly long and brilliant tail. There is little doubt that around 1303 when he came to that vital decision

*I am indebted to Roberta J. M. Olson's article in *Scientific American* (1979) for drawing my attention to this brief biblical reference.

about how to depict the 'Star in the East' he decided to depart from convention. He preferred to put it in as the blazing dynamic comet which he recalled so well from personal observation two years before rather than the geometrical, stylistic, somewhat naive, even clinical interpretations of previous illustrators in centuries past. So Giotto's innovation became the first realistic impression of the Comet and he has gone into history again, in perpetuity, by his name being given to the first major space mission to intercept the very Comet he saw 684 years earlier.

Let me add a curious coincidence in Giotto being commissioned for this work. His friend Dante, who also saw the Comet, depicted a wealthy businessman in Padua called Scrovegni as the great usurer when he wrote his famous *Inferno*. Enrico, Scrovegni's son, is said to have gained permission to rebuild on a Roman ruin the Arena Chapel which he hoped would atone for the bad reputation of his father, and not only did he ask Giotto to help in its construction but also to paint many frescoes depicting the life of Christ. The *Adoration of the Magi*, one of many frescoes, is the essence of the naturalism which enabled Giotto to exercise such authority on the style of subsequent great painters, whose work can be seen in Padua to this day.

There is no way, then, that Halley's Comet could coincide with the date of the birth of Jesus Christ. Its appearances in 12 or 11 BC and AD 66 leave far too great a gap, even allowing a tolerance of up to five years in determining the actual date of Christ's birth.

AD 66 Halley's Comet is observed by Chinese astronomers for about seven weeks. Josephus, the Jewish historian, sees Halley's Comet stretched out over Jerusalem 'like a sword'. The Old Testament (I Chronicles 21:16) says, 'And David lifted up his eyes and saw the angel of the Lord stand between the earth and the heaven, having a drawn sword in his hand stretched out over Jerusalem.' It may be describing the Comet.

A revolt against the Romans breaks out in Judaea. A Roman garrison is wiped out.

Jews and Gentiles fight each other. Gentiles massacre Jews at Caesarea.

In the adjacent years in chronological order we find that in Britain, Suetonius Paulinus, a Roman General, massacres the inhabitants of Anglesey.

In the South, Boudicca (Boadicea) and the Iceni are joined by the Trinovantes in a revolt against the Romans. Romans are massacred at London, Colchester and St Albans. Suetonius Paulinus comes south to deal with the rebel tribes. He defeats them in the Midlands. Boudicca commits suicide.

St Paul reaches Rome having survived a shipwreck off the coast of Malta. He is put under house arrest.

St Mark is by tradition martyred in Venice.

The great fire of Rome; the Emperor Nero is suspected of starting the fire but he uses the Christians as scapegoats and has many of them executed.

The Piso conspiracy to assassinate Nero fails and Seneca and Lucan are forced to commit suicide by Nero.

Titus and his father, Vespasian, are sent to Judaea to put down the Jewish revolt.

By tradition St Peter is crucified. St Paul is executed.

Roman legions revolt in Spain and Gaul. Nero flees from Rome under threat of his life. He commits suicide when he sees he cannot escape his enemies. Nero's death means the end of the Julio-Claudian line of Caesars.

Vespasian is now Emperor – the first of the Flavians.

After a six-month siege Jerusalem falls to the Romans under Titus. Titus burns the Temple and enslaves the survivors.

AD 141 Chinese astronomers report seeing an apparition in March and April which was Halley's Comet. A terrible plague sweeps many countries shortly afterwards with hundreds of thousands dying in Italy alone. The Comet is blamed even though the plague did not attack Europe until a long time after it had gone away.

In Britain Urbicus, the Roman governor, builds a wall from the Firth of Forth to the Firth of Clyde. It is thirty-seven miles long and becomes known as the Antonine Wall.

About this time the Christian, Justin of Samaria, conducts a defence of Christians before Emperor Antoninus.

Ptolemy, the astronomer, sees Halley's Comet from Alexandria.

In the preceding years, the Emperor Antoninus and the future Emperor, Marcus Aurelius, are the two Consuls. Juvenal, the Roman satirical poet, and Suetonius, the historian, die.

A Roman theatre is built at St Albans.

AD 218 A time of civil war in China.

At Rome, a power struggle develops between Macrinus and Julia Maesa, the sister of Julia Domna, who is the wife of the Emperor Severus.

Julia Maesa declares the Syrian Elagabulus, her fourteen-year-old grandson to be Emperor.

Macrinus and Julia Maesa meet in battle. Macrinus is defeated and executed.

Calixtus is deemed unworthy of being Pope by Hippolytus, a priest. Hippolytus goes about establishing a rival church.

Tertullian, a Roman convert to Christianity, turns to Montanism, a Christian movement that believes, contrary to Catholics, that Christians fallen from grace cannot be redeemed.

The Emperor Severus visits Britain and then dies at York.

In 217, Caracalla, son of Roman Emperor Severus, and who gave Roman citizenship to all free men in the Roman Empire, is killed by his own mutinying troops near to Edessa. The head of the Praetorian Guard, Macrinus, declares himself Emperor.

Macrinus banishes Julia Domna, the mother of Caracalla, to Antioch. There she starves herself to death.

In 219, the Emperor Elagabulus enjoys an extravagant and decadent way of living in Rome. His grandmother, Julia Maesa, takes political control while he holds banquets and revelries at enormous cost.

AD 295 Once again the appearance of Halley's Comet clearly recorded by Chinese astronomers.

In this and the adjacent years the Roman Empire is fighting for its life with intrigues and battles. It has been divided into the Western and Eastern Empires governed by two Emperors with two 'Caesars' who act as deputies. The two Emperors are now Diocletian and Maximian. Their Caesars are Galerius and Constantius Chlorus.

Constantius Chlorus, who is the father of Constantine the Great, prepares to invade Britain where there is a revolt led by Allectus. It was in this year that Allectus killed Carausius who had proclaimed himself 'Emperor of Britain'.

Galerius, the Caesar who is deputy to the Emperor Diocletian, is fighting against the Barbarians on the northern frontiers.

Constantius lands in Britain and meets Allectus in battle. Allectus is defeated and killed.

Constantius begins a rebuilding programme at London, York and St Albans. Hadrian's Wall is rebuilt. Britain's prosperity begins to revive.

Narses becomes King in Persia. He shows his readiness for fighting with an attack on the Armenian king, Tiridates, who is supported by Rome.

Marcellinus becomes the twenty-ninth Pope.

The Emperor Diocletian arrives in Egypt to deal with a rebellion that has broken out there. The future Emperor Constantine the Great is on Diocletian's staff.

The other Emperor, Maximian, reaches Carthage where he has to put down a rebellion.

Warring begins between Narses of Persia and the Romans under Galerius.

Only after a year's fighting do the Romans achieve victory and annex Persian territories.

AD 374 The trials of the Roman Empire continue, marked by more rebellions and the menace of the pagan tribes. In this year Theodosius, a Roman commander who earlier defeated the Picts and Scots when they invaded England, has just ended a rebellion in Morocco. His son, also called Theodosius, fights in the Quadi in Illyria.

St Jerome, doctor of the Church, goes to Syria and resides with hermits near Antioch. In the next year he will have a vision as a result of which he will denounce his pagan studies and concentrate on learning Hebrew to read the scriptures.

St Augustine is twenty years old.

In the three years immediately following the departure of the Comet, Valentinian, who became Emperor of the West in 364 dies. Gratian, his son, succeeds him.

St Ambrose is elected Bishop of Milan.

The Visigoths, who have been defeated by the Huns, ask the Romans if they can settle on Roman territory. The Romans mishandle the situation and in the end the Visigoths attack them.

The Battle of Adrianople: Valens, the Emperor of the East, is killed by the Visigoths. Gratian appoints his general, Theodosius, to succeed him.

Shortly afterwards, Hadrian's Wall is finally overrun and not rebuilt, and St Jerome translates the Bible into Latin.

AD 451 The Chinese record this as a spectacular appearance of the Comet lasting thirteen weeks. People in Europe believe it is the cause of the death of some 150,000 soldiers at the Battle of Chalons, when Attila and his Huns were defeated by the Romans under Aetius. This enormous decisive battle saved Gaul and Paris being conquered by the Huns. Attila however survived. See the story of the 'spirit battle' in Chapter 12 which legend has it that such was the ferocity of the fighting, the souls of the fallen continued the slaughter.

In the adjacent years Angles, Jutes and Saxons begin to conquer Britain. The Saxons, whose leaders are Hengest and Horsa, have come at the request of the Welsh King, Vortigern.

In AD 452 Attila the Hun invades Italy. The Emperor of the West, Valentinian III, makes a retreat to Rome while Pope Leo I is left to meet Attila to try to persuade him to come to a peace agreement. Pope Leo I manages to halt the Huns' advance.

Venice is founded in swampland by those who have fled from the Huns.

The following year, Attila the Hun dies. His empire is divided between his sons but the Huns no longer pose a serious threat to the West. Genserio the Vandal takes advantage of a decline in Roman power and attacks Rome. He captures and loots Rome.

St Patrick, the patron saint of Ireland, is now about seventy years of age.

AD 530 A terrible plague sweeps all over Europe.

Belisarius, the Byzantine general, defeats the Persians in northern Mesopotamia. He is then defeated at Callinicum. By this time the Roman Empire is in complete decline.

In the previous year St Benedict, who is about fifty years old, founds his monastery at Monte Cassino in Italy.

Justinian closes down the great academy at Athens founded by Plato a thousand years previously. Many Greek scholars migrate to Syria and Persia.

In AD 532 and 533 the building of the great Christian church, St Sophia, is begun at Constantinople.

Belisarius is sent by Justinian to North Africa to capture it from the Vandals.

The Franks capture the kingdom of Burgundy.

AD 607 Historical references are scant, except for one event affecting our world today. Mohammed has his vision of the Archangel Gabriel about this time. He is some forty years of age and in this decade the great spread of Islam takes root.

Parts of Europe are conquered by Slavic tribes. But the years on either side are not without their landmarks.

AD 604: Pope Gregory the Great dies. It was Gregory who sent missionaries, including St Augustine, to England, prompted, it is said, by seeing fair-haired Anglo-Saxon youths in a Roman slave market.

AD 606–7: Persian forces cross the Euphrates and head westwards.

The Avars capture territory around Constantinople.

AD 684 This is the Halley's Comet which was the first to be illustrated, but not the first illustration of the Comet, which comes into the Bayeux Tapestry (see pp. 74–5 and n.). The explanation is that a rough woodcut drawing of it was published in the Nuremberg Chronicle in 1493 just over 400 years after this Comet's appearance. The Chronicle recorded annual happenings and a primitive interpretation of the Comet accompanied the text on the page dealing with AD 684 and is indeed repeated as a space filler in other parts of the book. The text states that this year of the Comet brought deluges of rain, thunder, lightning and other acts of God causing animals and people to perish, followed by the withering of the harvest and the onslaught of plague.

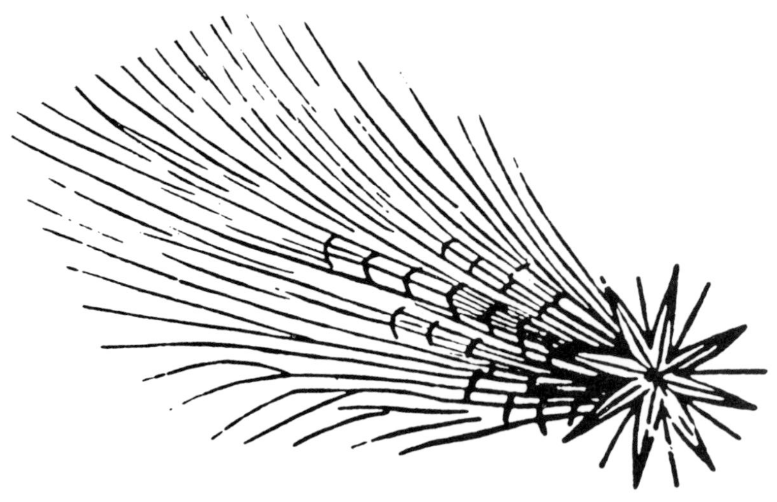

The spiky hedgehog impression of the first Halley's Comet to be drawn, that of AD 684 nearly 400 years before the Bayeux Tapestry, but not issued until 1493 as part of the narrative of events in AD 684 in the Nuremberg Chronicle.

AD 760 The dreadfully cold winter of AD 760–1 is blamed on the Comet even though its apparition had taken place in the spring before.

A Tartar tribe invades Armenia and sets the foundations of the Turkish Empire.

About this time the Book of Kells, the Gospels in Latin, was written. The book is now in Trinity College, Dublin.

In India, work begins on the Hindu Temple of Kailasha that is being carved out of rock.

In the years immediately preceding, Abd Al-Rahman founds an Arab dynasty in Spain. He makes Cordoba his capital.

Offa becomes King of Mercia. He will reign for thirty-nine years and construct an earthwork from Wales to Mercia that will be known as 'Offa's Dyke'.

Pepin, King of the Franks, drives out the Arabs from their settlements in France.

At about the same time Baghdad is founded by Al-Mansur. It will be capital of the Muslim world.

King Pepin the Short of the Franks dies. He is succeeded in 768 by his sons Charlemagne and Carolman. Carolman dies three years later and Charlemagne becomes sole King of the Franks.

AD 837 A number of comets are observed about this time, but Halley's was identified as having made probably its most spectacular return in history with a tail stretching right across the sky. It achieved its closest ever proximity to earth of about 6,000,000 kilometres or something under 4,000,000 miles, which is nine times nearer than it will get at its closest to earth in 1986.

The connection between the death of kings and the appearance of the Comet is reinforced. Shortly after it was sighted King Louis of France died. Events in Europe at the same time come more into prominence.

In Spain, a revolt by the people of Toledo is crushed by Abd Al-Rahman II, Emir of Spain.

Arab forces in Sicily come to the rescue of Andreas, Duke of Naples, who is being besieged at Naples by the Duke of Benevento.

Pietro Tradonico becomes Doge of Venice.

In the previous fourteen months Vikings again raid England and will soon cross the sea to ravage Ireland.

Egbert, the King of Wessex, who was recognised as overlord of England in 829, defeats the Britons and the Danes in Cornwall.

London is sacked by Vikings.

In AD 838–9 the Byzantine Emperor is defeated in battle by Al-Mutasim who decides against marching on Constantinople.

Egbert, King of Wessex, dies. His son, Ethelwulf, succeeds him.

AD 912 This visit of the Comet was also recorded by Japanese observers. Strife in England and Europe.

Battle of Tettenhall: Edward the Elder defeats the Danes. Halfdan, King of York, is killed.

King Louis the Child, who is eighteen years old, and the last Carolingian ruler of Germany, is defeated by the Magyars (who began their invasion six years before) in a battle near Augsburg.

William, Duke of Aquitaine, founds the Abbey of Cluny.

In the following two years history records the death of Louis the Child.

Ethelred of Mercia, dies. His wife, Ethelfleda, daughter of Alfred the Great, succeeds him.

A Byzantine fleet is defeated by the Arabs off the island of Chios, near the coast of Turkey.

The Byzantine Emperor, Leo VI, dies. His brother, Alexander, succeeds him.

Rolf the Ranger, who has been baptised a Christian, settles in Normandy with his Norsemen.

Work on the great earthen mound of Warwick Castle is begun.

AD 989 The Comet is described in European as well as Chinese reports. It was noted by the Saxon historian, Elmacin.

Viking attacks occur in the south-west of England.

Archbishop Aldabero of Reims dies. Arnulf, a bastard son of Lothair, the late King of the Franks, succeeds him.

Basil II, the Byzantine Emperor, and Bardas Phocas, the Pretender to the throne, meet in battle near Abydus. Phocas is defeated.

Smbat II is succeeded as King of Armenia by his brother Gakik I.

In the previous year, the Byzantine Emperor, Basil II, baptises Vladimir, Grand Duke of Kiev, in Constantinople. On returning home, Vladimir begins the conversion of the Russians to Eastern Orthodoxy.

A Byzantine army with help from Vladimir of Kiev defeats Bardas Phocas, the Pretender to the Imperial throne.

St Dunstan dies at Canterbury. He is credited with the revival of English monasticism which had petered out with the Viking invasions.

The stylised 'shuttlecock' version of the Bayeux Tapestry's depiction of the 1066 Comet. The earliest of all the drawings, but *not* the first apparition to be drawn.

1066 Famous in English and European history. The Comet appears and William of Normandy looks upon it as a favourable omen, as 'a wonderful sign from Heaven', and urges his soldiers forward in battle with it. The order of events that fateful year were briefly:

Jan 5th: Edward the Confessor, King of England, dies.

Jan 6th: Harold is elected King.

Sept 25th: Battle of Stamford Bridge. Harold defeats Harold Hardrada, King of Norway, and Tostig, his outlawed brother.

Sept 28th: William of Normandy invades England.

Oct 14th: Battle of Hastings. King Harold has come south to fight William of Normandy. Harold's men, stronger in numbers but inferior in quality, resist the Normans till sunset. William then orders his archers to fire high in the air so that the arrows drop down upon the enemy. The plan is effective. Legend has it that Harold is hit in the eye and the battle is lost for the English.

The famous Bayeux Tapestry*, begun in the next year, commemorates the Norman victory at Hastings and the appearance of Halley's Comet, earlier in the year.

Dec 25th: William the Conqueror is crowned King of England.

*When I visited Bayeux to see the famous Tapestry in September 1984 to check the facts for this book, I was struck by a number of things. There was no particular prominence given in the texts explaining the significance and origins of the Tapestry, nor, among the proliferation of postcards and booklets and other souvenirs, of Halley's Comet and its imminent return. It is quite rightly considered to be 'the longest cartoon strip in the world', being no less than nearly seventy metres in length, made up of eight pieces of cloth ranging from just over five metres to nearly fourteen metres and each section being only half a metre in width.

The reason it was so long and narrow was because it was designed simply as a hanging made of material embroidered with pictures and captions in Latin representing the conquest of England to be stretched around the nave of Bayeux Cathedral on special occasions and on feast days. It was hung from pillar to pillar as a sort of decorative banner to give the somewhat illiterate faithful a pictorial history of William's great and justifiable victory. I use the word 'justifiable' because the story line emphasises that Harold was the 'baddie'. It explains that King Edward of England asked his brother-in-law Harold Godwinson to go to France to tell William that as Duke of Normandy and Edward's cousin he had been chosen to succeed Edward. On the way Harold is blown off course and falls into the hostile hands of Count Ponthieu who demands an enormous ransom for his release. William pays it, is very kind to Harold, and Harold is then depicted as taking a most sacred oath over the bones of saints and a holy ampulla containing a drop of Christ's blood collected during the Crucifixion. The inference is that Harold had sworn a most solemn and binding oath to recognise William as successor to the English throne when Edward dies. Harold is seen off safely to England, laden with good wishes and gifts and then what does he do? As soon as Edward dies he grabs the throne for himself. So no wonder William took it upon himself to exact retribution for the breaking of this most sacred oath and invaded England. The rest is history. Harold, the 'baddie' in the Norman version, simply had to lose.

Finally it is interesting to note two things. The first is the likelihood that it was Odo, a half-brother of William, who was appointed by William to be the Bishop of Bayeux, who organised the making of the Tapestry and not Queen Matilda, William's wife, as legend has it. Secondly, if you think hard 'porn' was not invented in pictorial fashion until the twentieth century, you should look at a few of the small panels which edge the whole length of the Tapestry at top and bottom. Close inspection of Sections 13 and 15 reveal these fascinating drawings. Remember the Tapestry was completed and first exhibited on July 14th (coincidentally the Bastille Day of 1789) in 1077, and that is a long time before today's pornographic printed matter arrived.

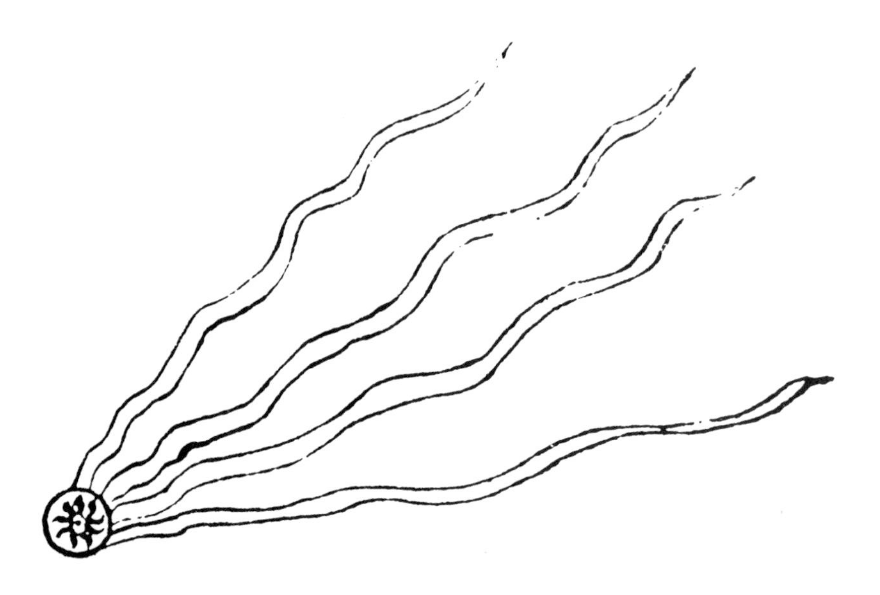

The horizontal jellyfish in the Canterbury Psalter doodled by the Monk Eadwine in his copy of an illuminated manuscript of the Psalms when he saw the Comet of 1145.

1145 There is what is thought to be a drawing of the Comet, looking like a horizontal jellyfish, in the Eadwine Psalter (Canterbury), an illuminated manuscript by a monk of that name who copied the Psalms from the earlier Utrecht Book of Psalms. Above the drawing are three Latin versions of the fifth Psalm which appear to have no relevance, but the caption or legend accompanying the Comet cites the brightness of the 'hairy star' and says that comets are rare and appear as portents.

Up till now the Comet has been both named as a portent and also blamed for many things on its periodic apparitions, but this time it is particularly invoked in the name of Christianity. Pope Eugenius III is inspired by the sight of it to call for a Holy War against the Muslims. He proclaims the Second Crusade.

Geoffrey of Anjou, the husband of Matilda, is now the Duke of Normandy.

Robert of Chester makes the first Latin translation of the *Algebra* of Al-Khwarizmi.

The towers of Chartres Cathedral are being built.

A year or so later the Second Crusade begins: Conrad III of Germany and Louis VII of France lead the Christian forces.

Matilda departs from England in 1148, no longer contesting the English throne, and leaves the way clear for Stephen.

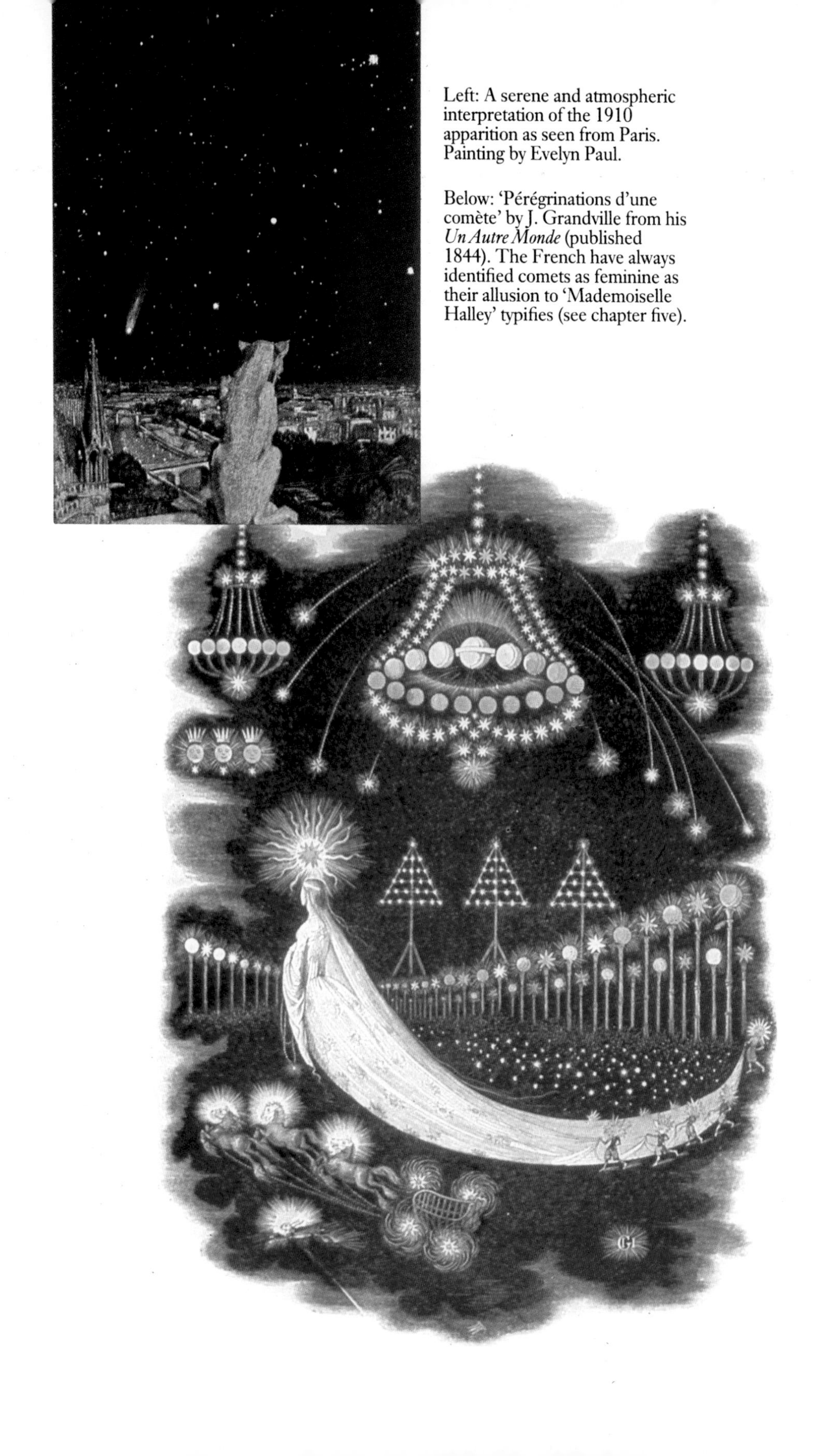

Left: A serene and atmospheric interpretation of the 1910 apparition as seen from Paris. Painting by Evelyn Paul.

Below: 'Pérégrinations d'une comète' by J. Grandville from his *Un Autre Monde* (published 1844). The French have always identified comets as feminine as their allusion to 'Mademoiselle Halley' typifies (see chapter five).

Part of the collection of Babylonian tablets in the British Museum which offer the first confirmation of the 164 BC comet and additional evidence to supplement the Chinese reference to the 87 BC comet. (See chapter seven).

1222 This is the time of the fearsome Genghis Khan who believed that comets were his special stars, and when Halley's arrived he sent in his Mongolian hordes to slaughter thousands of people when he destroyed Herat in north-west Afghanistan.

In the same year in England, the Synod of Oxford establish St George's Day on April 23rd.

The University of Padua is founded.

During the two adjacent years, both before and after, Genghis Khan destroys Bukhara.

The Fifth Crusade ends.

The building of Salisbury Cathedral is begun.

St Francis of Assisi steps down from the administration of the Order he founded.

Genghis Khan invades India.

St Dominic, founder of the Dominican Order, dies. Dominican friars land in England.

Genghis Khan invades China.

War breaks out between England and France.

Franciscan friars arrive in England.

In 1227, the year St Thomas Aquinas is born, Genghis Khan dies. His Empire which stretched from the Caspian Sea to the Pacific Ocean is divided up.

1301 No surprises on this occasion for the Comet. The main events could be covered by three words, 'War in Europe'. Too numerous to itemise, except perhaps, for the Battle of Baphaion, when Osman I, who founded the Ottoman dynasty, defeats the Byzantines. Worth noting is that in the same year:

Edward I is the King of England. He is the eldest son of Henry III and Eleanor of Aquitaine. It was during his reign that the founding of the English parliamentary system and the organisation of the Law Courts occurred.

Edward I confers the title of Prince of Wales on his son Edward.

About this time, Marco Polo, the Venetian traveller, was in a Genoese prison writing, with the help of a fellow prisoner, an account of his travels.

The two Florentines, Dante and Giotto, would have seen Halley's Comet this year. Giotto will use the Comet to represent the Star of Bethlehem in his picture of the Nativity.

1378 Halley's Comet seen in China and Europe.

A year in which Pope Gregory XI dies, and when the start of the Great Schism takes place. As a result of a proclamation by Pope Urban VI saying that he will reform the Church, a number of cardinals elect another Pope – Pope Clement VII – who is based in Avignon in France.

A tract by John Wycliffe urges that 'Peter's pence' should not be paid to Rome. He gets the support of John of Gaunt.

Geoffrey Chaucer, the father of English poetry, is about thirty-eight years old and is yet to write his great poems *Troilus and Criseyde* and the *Canterbury Tales*.

In the previous year Edward III, whose reign was marked by victories over the French and the Scots, dies. Richard II, Edward's grandson, is the new King.

John Wycliffe (mentioned above) is accused of heresy. Wycliffe was an English cleric who challenged the authority of the papacy and demanded Bibles written in the common language of the people.

In the year after the Comet's visit, Thomas à Kempis, author of the classic spiritual work *The Imitation of Christ*, is born.

William Wykeham founds New College, Oxford.

A little later St Catherine of Siena dies at the age of thirty-three. She was an important mystic and spiritual writer and received the stigmata.

In England, peasants led by Wat Tyler revolt and march on London where they confront the young King Richard II. Tyler is betrayed and killed.

1456 This time the Comet witnesses religious strife yet again. There is a great war between Muslims and Christians. Pope Callixtus III is said to have 'excommunicated' the Comet, and offers up prayers for deliverance from it and from the Muslims who are besieging Belgrade. The Pope certainly did order the ringing of church bells 'to aid by their prayers all those engaged in battle with the Turk'. Both sides are frightened of the Comet.

Paolo Toscanelli, observing the Comet from Florence, stated: 'Its head was round and as large as the eye of an ox, and from it issued a tail, fan shaped like that of a peacock. Its trail was prodigious, for it trailed through one third of the firmament.'

In the same year the Turks capture all of Athens except for the Acropolis which holds out for two more years.

The trial of Joan of Arc, which happened in 1430–1, is declared to be irregular by a French ecclesiastical court.

The Cape Verde Islands are discovered. They will be visited by Halley on his Atlantic voyages at the end of the seventeenth century.

1531 Halley suspected that this Comet and the one that appeared in 1607 were the same as the one he observed in 1682.

In this year Pope Clement warns Henry VIII not to take another wife until his divorce petition has been considered.

Earthquake at Lisbon.

The Spaniard, Francisco Pizarro, crosses the Andes.

The Inquisition is set up in Portugal.

In England, statutes are passed ordering unlicensed beggars to be whipped.

1607 Although this was the last return before the telescope was introduced, the observation of the Comet was getting more scientific with Kepler attempting to measure its track. One European astronomer observed that at its best it was like a 'flaming sword'. (See the AD 66 appearance.)

In England, peasant riots break out in the Midlands over the enclosure of land.

Union of England and Scotland rejected by the English Parliament.

English settlers facing great deprivation in North America found Jamestown, Virginia.

Literature: *Volpone* by Ben Jonson; *Timon of Athens* by Shakespeare.

In the two previous years there is the Gunpowder Plot, a Catholic plot to blow up the Houses of Parliament which fails. Guy Fawkes is arrested. He and his fellow conspirators are executed.

Literature: William Shakespeare's *Macbeth* and *King Lear*.

In the two years following 1607 the French establish a trading post at Quebec.

The great English poet, John Milton, is born.

Johannes Kepler sets out his first two laws of planetary motion, namely that planets move round the sun in elongated circular or elliptical paths and that they do not move at a uniform speed.

1682 Halley sees the Comet which will bear his name and studies it. He concludes it is the one that appeared in 1531, and 1607, and predicts later that it will reappear 'about 1758'.

Louis XIV is recognised as supreme ruler in France. His court is based at the Palace of Versailles which he is extending and improving.

Work begins on the Chelsea Hospital for disabled soldiers designed by Sir Christopher Wren.

The Earl of Shaftesbury is charged with treason. He flees the country.

The Ashmolean Museum in Oxford is founded.

Halley returns to England from his tour of the Continent and later on in the year marries Mary Tooke.

In the following three years the Rye House Plot, a plot to kill Charles II and his brother James, fails.

Purcell is appointed composer to King Charles II.

Halley begins his regular observations of the moon.

Robert Hooke invents the heliograph.

In the coldest winter in memory (1684) the Thames freezes over.

Halley visits Newton for the first time and the two men discuss the problem of planetary motion.

Charles II dies. His brother James II becomes King (1685).

Judge Jeffreys conducts his 'Bloody Assizes' on Monmouth and his supporters after a failed rebellion.

Handel and J. S. Bach are born.

Halley is appointed Clerk of the Royal Society.

1759 The Comet that Halley predicted would return more than fifty-five years before is detected on Christmas Day 1758. It will reach perihelion in 1759. It is the first time that the visit of a comet has been successfully predicted and it is named after Halley.

General Wolfe takes Quebec from the French. Both Wolfe and the French General Montcalm are killed.

Jesuits are expelled from Portugal.

The British Museum is opened at Montagu House. The move to its present site takes place in 1823.

Kew Gardens are opened.

Robert Burns, William Wilberforce and Horatio Nelson are born.

Literature: Voltaire's *Candide* and Dr Johnson's *Rasselas*.

Some of the events before and after this astronomical landmark are given in chronological order.

1755: An earthquake in Lisbon kills 30,000 people.

Dr Johnson publishes his famous dictionary of the English language.

1756: The Centenary of Halley's birth.

The Seven Years' War: Frederick II of Prussia invades Saxony.

The Black Hole of Calcutta: 123 English soldiers are alleged to have suffocated to death, having been imprisoned in a tiny room by the Nawab of Bengal.

Mozart is born.

1757: Frederick II wins battles over the French and Austrians at Rossbach and Leuthen.

William Blake and Thomas Telford are born.

1760: George II dies. His twenty-two-year-old grandson George III succeeds him.

Robert Clive returns from India.

1761: William Pitt resigns as Secretary of State for War.

In England, the Bridgewater Canal is opened.

1835 The Great Trek begins: from about this time, some 10,000 Boer farmers leave the Cape of Good Hope to rid themselves of British domination. They found Natal, Transvaal and the Orange Free State.

Lord Melbourne becomes Prime Minister after the resignation of Robert Peel.

The Municipal Corporation Act: local governments are reorganised.

Madame Tussaud's Waxworks are opened in London.

The US inventor, Sam Colt, patents his revolver.

Literature: *Fairy Tales* by Hans Christian Andersen; *Paracelsus* by Robert Browning.

Mark Twain, the American writer, is born. He will die in 1910 when Halley's Comet returns.

Some of the events in the adjacent years are set out, beginning with a repetitive and ominous item in 1831, and ending on an ironic note in 1838.

1831: The Russians suppress a revolt in Poland.

Charles Darwin sets out on a scientific expedition on board the *Beagle*.

Balzac, Victor Hugo and Stendhal are all writing in France.

1832: In Britain the first parliamentary Reform Act designed to change the electoral system is passed.

The great German writer, Goethe, author of *Faust*, dies.

1833: The Falklands are proclaimed to be a Crown Colony by Britain.

Johannes Brahms and Alfred Nobel are born.

1834: The Houses of Parliament are almost totally destroyed by fire.

The Tolpuddle Martyrs: six labourers try to form a union and are condemned to transportation.

1836: Adelaide is founded in Australia.

Charles Dickens publishes *The Pickwick Papers*.

1837: William IV dies. Queen Victoria comes to the throne.

In the USA Samuel Morse develops the telegraph.

1838: British troops invade Afghanistan to stop Russia's influence from spreading and to protect British interests in India.

1910 A detailed description of the public reaction has been given in a previous chapter but briefly points of interest this year are: Edward VII dies and George V becomes King.

Mark Twain, who was born in 1835 when Halley's Comet last appeared, dies.

Florence Nightingale and Tolstoy, author of *War and Peace*, die.

The *Daily Mail* offers £10,000 for the winner of an air race from London to Manchester. Won by a Frenchman, M. Paulhan.

Knossos, the ancient Minoan town in Crete, is excavated by Sir Arthur Evans.

Stravinsky's *Firebird* is performed.

Bertrand Russell and A. N. Whitehead publish *Principia Mathematica*.

Mme Curie experiments with radium.

1908: Baden-Powell founds the Boy Scouts.

Asquith becomes Prime Minister.

1909: Shackleton returns from the Antarctic, having failed to reach the Pole.

The University of Bristol and Queen's College, Belfast, are founded.

1911: Roald Amundsen is the first man to reach the South Pole.

Lord Rutherford develops his theory of atomic structure.

Britain's population is 40.8 million people.

1912: SS *Titanic* sinks after collision with iceberg.

1913: Stravinsky's *Rite of Spring* is performed.

Literature: *Sons and Lovers* by D. H. Lawrence; *Death in Venice* by Thomas Mann.

Richard Nixon, US President 1969–74, is born

Tycho Brahe (1546–1601). The Danish astronomer who proved that comets are not part of the earth's upper atmosphere.

8

Astronomy up to the Time of Halley

Astronomy between the sixteenth century and Halley's youth in the seventeenth century had undergone the most exciting and radical change since the start of the Western astronomical tradition that began with the ancient Greeks.

From the second century AD up until the sixteenth century the view of the universe that prevailed was the one described by Ptolemy in his book the *Almagest*, a grand synthesis of Greek astronomical views.

The Ptolemaic system placed the earth at the centre of the universe with the sun, moon and planets revolving round it, attached to solid, transparent, crystal spheres. Beyond the last planet, Saturn, lay the sphere to which the stars were fixed in place, providing a backdrop for the other celestial bodies that moved in perfect circular orbits.

The Ptolemaic system became interwoven with the authoritative views of the Greek philosopher Aristotle and medieval Christianity. It also conveyed beauty and harmony with its crystal spheres and perfect circular orbits.

It is no wonder that when Copernicus published his famous book *De Revolutionibus* in 1543, proposing that the sun was at the centre of the universe, the reaction was, when the implications had been fully realised, extremely hostile. This sun-centred or heliocentric theory was too much for religious orthodoxy. In 1616 Copernicus's ideas were outlawed by the Catholic Church. In 1632 Galileo was forced by the Church to recant his Copernican beliefs.

In England at this time, the Church's opposition to Copernicus was not very evident. Thirteen years after Copernicus's death, two books affecting his ideas were published in London.

The first came from the great Danish astronomer Tycho Brahe. Tycho was able to show that the comet that appeared in 1577 lay beyond the moon. This was startling because it had been thought up till then that comets were a phenomenon of the upper part of the earth's atmosphere. But since Tycho's observations showed that this comet was moving amongst the planets, it meant that the comet would have to have smashed through the solid, crystal spheres. Quite clearly it had not.

When Tycho Brahe died in 1601, Johannes Kepler (who had worked with him) made full use of his highly accurate astronomical observations and showed that the planet Mars had an elliptical orbit, not a circular one. Later, in his three famous laws of planetary motion, Kepler stated that all planets moved in elliptical orbits. The aesthetically pleasing 'perfect' circle inherited from the Greeks had now been quashed.

Galileo then furthered the astronomical revolution with his newly built telescope, through which he looked at the moon. To his and everyone else's shock and horror the moon seemed barren and cratered and not made of some shining, eternal substance, as had been believed.

Furthermore, Galileo answered the objection to the Copernican theory that if the earth moved round the sun, why did it not leave the moon behind? Galileo spotted four of Jupiter's satellites and pointed out that, since Jupiter did not leave behind its satellites, there was no reason why the earth should leave behind the moon.

The new astronomical ideas and discoveries of the Copernican revolution were later integrated in another way by the great French philosopher René Descartes ('I think, therefore I am').

In his book, *Principia Philosophiae* (published in 1644, twelve years before the birth of Halley) Descartes put forward his complex theory that the cosmos was full of whirlpools of matter.

Galileo (1564–1642). The versatile Italian scientist who built a telescope with which he saw that the moon was cratered and not the smooth disc everyone had previously assumed.

These vortices had fiery particles and were supposed to condense and form a number of stars. The central star within each vortex then became covered over with thicker particles until its light was completely shut off. Because of the consequent lack of outward pressure exerted by the star, it collapsed and formed a planet which would then stray off and be captured by another vortex, whose central star it would then orbit.

Descartes's vortex theory may seem far-fetched, but worked out in full it did make a bold attempt to explain various natural phenomena in a logical way. Certainly Halley would have been intrigued. At the time he went up to Oxford Descartes's theories were being taught in the English universities.

It is during the lifetime of Halley and his brilliant contemporary, Isaac Newton, that we shall see astronomy take further bold steps into the future.

Copernicus (1473–1543), whose proposition in 1543 that the sun not the earth was the centre of the universe shocked religious orthodoxy; his ideas were outlawed by the Catholic Church.

Flamsteed (1646–1719). First Astronomer Royal, appointed by King Charles II in 1675. His hostility to Halley is fully described in Chapter 9.

9

The Early Life and Career of Edmond Halley

Edmond Halley was 'of middle stature, inclining to tallness, of a thin habit of body, and a fair complexion, and always spoke as well as acted with an uncommon degree of sprightliness and vivacity', according to the *Biographica Britannica*, published fifteen years after Halley's death in 1742.

The *Biographica Britannica* also quotes evidence of Halley's character given by one of his contemporaries:

> The reputation of others gave him no uneasiness, and a restless jealousy and anxious emulation were strangers to his breast. He was equally ignorant of those extravagant prejudices in favour of one nation which are injurious to all others. The friend, countryman and disciple of Newton, he spoke of Descartes with respect.

Later it adds that he had

> a vein of gaiety and good humour which neither his abstracted speculations, the infirmities of old age, nor the palsy itself which seized him some years before his death, could impair, and this happy disposition, the gift of Nature, was the more perfect as it was still attendant upon that peace of mind which is the nobler endowment of virtue.

Even allowing for the exaggerations of contemporary accounts, Halley seems to have been a remarkable man; he was modest, charming, affable, intelligent, sensitive to the feelings of others, courteous, confident, responsible, hard-working, just in his dealings, open to new ideas and sweet-tempered.

Had Halley not lived at the same time as Sir Isaac Newton, he would have been the greatest English scientist of his generation. For, although he is now famous only in connection with his Comet, he was a brilliant and versatile astronomer and scientist, whose interests lay in many different fields including navigation, magnetism, classical studies and cosmology.

Edmond Halley was born on October 29th, 1656 at Haggerston in the East End of London which was, at that time, rural. His parents had married only seven weeks before his birth.*

*Too much should not be read into this: in 1653 an Act of Parliament was passed by which civil marriages before a magistrate were permitted, after which the married couple could then have a church ceremony if they wished. It might have been that Halley's parents had been married for some time before the church marriage.

Halley's father, who was also called Edmond, was both a soapboiler (a manufacturer of soap) and a salter. He also seems to have had an income from property rented out in the city. Halley had a brother, Humphrey, who died in his youth, and a sister, Katherine, who died in infancy. Not much is known about Halley's mother, even the date of her death, though she was buried on October 24th, 1672, five days before Halley's sixteenth birthday.

Very little is known of how Halley spent his early youth. We do know that he went to St Paul's School, where Milton and Samuel Pepys had been pupils. The buildings of the school burned down in the Great Fire of London in 1666, and would have been being rebuilt at the time when Halley was there, though the exact dates of his attendance are not known.

At St Paul's Halley would have studied the traditional subjects of that time, which were Latin and Greek and mathematics. Halley would also have been taught the practical application of mathematics in astronomy and navigation. Halley himself tells us that he was keenly interested in astronomy from his 'tenderest youth'.

In 1671, at the age of fifteen, Halley became Captain of the school, which suggests that his intelligence was matched by a sense of responsibility and leadership – qualities that he would exhibit later on in his life, especially during two scientific voyages he made across the Atlantic.

In 1672, the year the third Anglo-Dutch war broke out and his mother died, Halley made a measurement of the variation of the magnetic compass, which was his first scientific observation on record.

In 1673, Halley went up to Queen's College, Oxford, the university where he would later become a professor. By this time, he had had a solid grounding not only in Latin and Greek but also Hebrew, whose value lay in connection with Bible studies. To encourage his son in his pursuits, Halley's father very generously bought him a number of astronomical instruments, including a twenty-four-foot-long telescope and a sextant. Halley was now set up for a flying start in the world of astronomy.

At Oxford, he was lucky enough to come under the influence of two distinguished professors: one was the mathematician, the Reverend John Wallis (1616–1703), who had decoded captured Royalist letters during the English Civil War, and the astronomer Edward Bernard (1638–96).

During his student days, Halley carried out observations with his telescope and in 1675 he was able to write to none other than John Flamsteed, later to become the first Astronomer Royal, and tell him of observations he had made of an eclipse of the moon. This indicates the degree of confidence Halley had in his own ability. He also told Flamsteed that he had found errors in the catalogue of stars published by the famous Danish astronomer Tycho Brahe (1546–1601).

Flamsteed, with whom Halley was to have close professional links for the next forty years, not always with happy results, was born in 1646 and had been appointed the first Astronomer Royal by Charles II in March 1675. In the following year, Flamsteed took up residence at Greenwich where the new Royal Observatory had been built to rival the French Observatory at Paris. When Flamsteed died in 1719, Halley succeeded him as Astronomer Royal.

In 1675, Halley started work on a paper on the motion of the planets, developing the work of the great German astronomer Johannes Kepler (1571–1630), who had shown that planets did not move in circular orbits round the sun, as had been thought until his time, but that they moved in 'elliptical' (elongated circular) orbits. It was Halley's first scientific paper and was published in *Philosophical Transactions*, the journal of the Royal Society, the premier scientific society in Britain. At the age of nineteen Halley had made a bold and intelligent start to his career.

Halley left Oxford at the age of twenty, before taking his degree, in order to go to the island of St Helena to chart the stars of the southern hemisphere. Why he left before taking a degree is not known. The most likely answer is impatience – a young man's desire to get on with his career.

The stars act as points of reference for astronomers and navigators and so it is essential to have an accurate chart of their positions in the sky. Such able astronomers as the Italian Jean Dominique Cassini (1625–1712) then in Paris, Johannes Hevelius (1611–87) in Danzig and John Flamsteed in London were concentrating their efforts on the stars of the northern skies. To make a catalogue of the southern stars must have seemed an ideal way for a young astronomer to make his name.

The island of St Helena is a Crown Colony in the South Atlantic Ocean off the west coast of Africa. Discovered by the Portuguese in 1502, eventually it was annexed by the East India Company, the powerful British trading company, in 1659. Napoleon was kept prisoner on St Helena from 1815 till his death in 1821. It is a volcanic island about seventy-five square miles in area and far enough into the southern hemisphere to have enabled Halley to chart the relatively unknown southern stars.

Halley needed to get permission from the government and the King to go to St Helena, and also from the East India Company. He was able to mobilise support for his cause from John Flamsteed and Henry Oldenburg, an Honorary Secretary of the Royal Society. The King gave his blessing without any delay and so did the East India Company, aware, perhaps, that the trip might bring advances in navigation that they could profitably use.

In February 1677, the expedition landed on the island and Halley and his assistant, Mr Clarke, set up an observatory on a slope that was later to overlook Napoleon's tomb. Contrary to expectation, the weather was bad and observations of the stars were hard to come by. In late November Halley wrote to Sir Jonas Moore, one of the patrons of his trip, and told him that

> the Horizon of this Island is almost always covered with a Cloud, which sometimes for some weeks together hath hid the Stars from us, and when it is clear, is of so small continuance, that we cannot take any number of Observations at once; so that now, when I expected to be returning, I have not finished above half my intended work; and almost despair to accomplish what you ought to expect from me.

In addition, matters were made worse by the behaviour of the governor of the island, Mr Gregory Field. Mr Field treated Halley with the utmost incivility. Later he became such an embarrassment to the East India Company that he was dismissed from his post in early 1678.

Despite problems of the weather and Mr Field, Halley was able to chart the positions of 341 stars, which was the primary aim of the trip. He also made other observations and discoveries, notably that, in order to keep correct time, the pendulum of his clock had to be made longer than it was in England, a phenomenon now known to be due to the equatorial bulge of the earth. He also observed the 'transit' or passage of the planet Mercury across the disc of the sun. This was important, because it gave Halley the idea of using observations of the transit of another planet, Venus, across the sun as a means of calculating the distance of the sun from the earth.

In May 1678, Halley returned to England and his scientific results were published as *A Catalogue of Southern Stars*. It was the first publication of the positions of stars that had been observed telescopically, rather than with the naked eye. Apart from other astronomical information – such as the transit of Mercury across the sun – Halley mentions in the book how he thinks the moon might be useful for calculating longitude at sea.

The energy invested in astronomy at this time was partly due to the needs of navigators, who relied upon knowledge of the positions of celestial bodies by which to steer their ships. In the late seventeenth century, overseas trading had expanded rapidly and was a source of fierce rivalry between the English, the Dutch, the Portuguese and the French. To know exactly where his ship was on the ocean, a navigator had to find out his latitude (distance above or below the equator) and his longitude (distance east or west of a standard meridian, such as the line that goes from north to south through Greenwich). To find his latitude was relatively simple. An instrument (a sextant) was needed for measuring the altitude of the sun in conjunction with a book of tables of the apparent annual motion of the sun through the sky. To find longitude was much more difficult, requiring a comparison between the ship's local time with a standard time (as Greenwich Mean Time is). The clocks in Halley's day were not accurate enough to be used for this purpose, so it was suggested that the moon could be used as a sort of giant clock hand pointing out standard time against the backdrop of stars. But the moon's motion through the sky is extremely irregular, and a pattern can only be discerned over a period of eighteen years. Halley would eventually undertake the task of observing the moon through an eighteen-year period in a bid to solve the problem of longitude when he became Astronomer Royal in 1720. It is, however, interesting to note that even at the age of twenty-one, Halley had given thought to this important problem.

Recalling a time when many ships left port never to return because of their inability to identify their position accurately, Dr Stuart Malin, Head of the Department of Astronomy and Navigation at the Old Royal Observatory at Greenwich, pinpointed the need for its solution as the equivalent today of finding a cure for cancer. This he did in a letter to the Dean of Westminster Abbey in July 1984 supporting the idea that Halley deserved a niche for himself there.

A Catalogue of Southern Stars was published in November 1678. Robert Hooke (1634–1703) paid it the honour of bringing it before the Royal Society, and in the following year it was translated into French. Halley presented a copy of his planisphere (a chart depicting stars and constellations) to Charles II as an expression of gratitude for the King's support of the expedition. Also, in honour of the King,

A newspaper boy looking through a street telescope in the United States. Drawing by Sidney Riesenberg from *Harper's Weekly*, May 1910.

The 1910 equivalent of a television 'soap' opera was the thrilling newspaper serial. The *Chicago Ledger's* 'In the Comet's Track: or Strange Adventures in Space' was typical of the comet-orientated serials of the period.

An imaginative illustration of a comet as part of the W.D. and H.O. Wills cigarette card set in 1928 dealing with the heavens.

At least seventeen compositions dedicated to Halley's Comet were published in 1910. They ranged from marches and two-steps, to gavottes and rags. None appears to have survived in the general musical repetoire, but these four illustrations of the sheet music covers convey the ambience of the period.

Halley added a new constellation to the traditional ones, calling it *Robur Carolinum*, 'The Oak of Charles', referring to the tree in which the King hid after the Battle of Worcester.

Halley needed royal influence to obtain the degree he had failed to get at Oxford owing to his premature departure. In due course, he was awarded his Master of Arts in December 1678. He was also informally awarded by Flamsteed the title 'Southern Tycho' after the renowned Danish astronomer Tycho Brahe.

In 1679, Halley made an important visit to the German astronomer Johannes Hevelius (1611–87), who lived in Danzig (modern-day Gdansk). The trip had the blessing of the Royal Society, because it was anxious to resolve a controversy that had arisen between Hevelius and the Englishmen Flamsteed and Robert Hooke. The controversy revolved around the type of sights used on observational instruments. In short, Hevelius favoured open sights (that resembled rifle sights), while Flamsteed and Hooke were convinced of the superiority of telescopic sights, which were systematically replacing open sights by the 1670s.

The proponents of telescopic sights maintained that it did not matter how sophisticated the observational instrument was – if it had open sights, it was inevitably limited to the power of the naked eye. On the other hand, Hevelius pointed out that telescopic sights had the disadvantage of being prone to unwanted optical effects. In principle the Englishmen were right, since the magnifying power of telescopic sights simply does not exist with open ones. But at the time it would be correct to say that the making of telescopic sights was still at an early stage of development and that Hevelius had a fair point about their potential unreliability.

The controversy became more intense after the publication of the first section of Hevelius's famous work *Machina Coelestis* (literally 'Machine of the Heavens' or 'Celestial Mechanics') in 1673. It was then revealed that he had used open sights on all his instruments. Flamsteed and, in particular, Hooke openly made it known that they were doubtful of the reliability of Hevelius's findings. Not surprisingly, Hevelius, who was an extremely meticulous, highly skilled and experienced astronomer, was upset by these assertions, especially since Flamsteed's own observational results, made with telescopic sights, were no more accurate than his own. The way Hevelius must have seen it was that his arduously accomplished work was being doubted on the basis of a theoretical superiority that had no concrete evidence to support it.

Henry Oldenburg of the Royal Society (to which Hevelius had been elected Fellow in 1664) tried to intervene and calm the situation between the two sides. It was at this moment that Halley entered the arena by writing a courteous letter to Hevelius, probably encouraged to do so by the Royal Society who disowned Hooke's criticism of Hevelius and wanted to reconcile the two sides. Halley sent Hevelius a copy of his newly published catalogue of the southern stars, and ended his letter by saying that he, Halley, intended to visit him at Danzig to observe and try to understand his astronomical methods. This demonstrates the confidence and initiative of Halley in wanting to meet and stay with one of the most renowned astronomers in Europe. Halley was the perfect person to mediate between Hooke and Hevelius, because he was an able enough astronomer to understand the merits of the respective claims as well as having considerable diplomatic skills.

Halley arrived in Danzig in May 1679 and stayed there two months. After ten days he wrote to Flamsteed and told him about Hevelius's instruments, especially the six-foot sextant that needed two people (one of whom did not require astronomical expertise) to operate it. More important, Halley testified to the consistently high degree of accuracy that Hevelius was able to achieve with his methods.

It is not surprising that Hevelius found Halley a delightful guest who was both interested in everything he saw and extremely gracious. Quite sensibly, Hevelius asked Halley to write, before his departure from Danzig, a testimonial endorsing both the instruments and the results that he had seen for himself. This Halley willingly did, and Hevelius published the testimonial along with their joint observations in his book called *Annus Climactericus* which was published six years later. It must have seemed to Hevelius that, for the present at least, he and his methods had been vindicated.

When the *Annus Climactericus* was first read in England in 1685, a number of small errors were found in Hevelius's text; for example, he did not give credit to Halley for going to St Helena on his own initiative and he mistakenly referred to Halley's quadrant as a sextant. Then, perhaps because of his advancing years and a residual bitterness over the whole open/telescopic sights controversy, Hevelius claimed that Halley had been sent to Danzig with the sole aim of spying on him. He also alleged that Halley had been sent on the scientific expedition to St Helena specifically at Hevelius's bidding. As a result of these wild claims, Halley, who had been a friend and ally up till then, referred to him in a letter to William Molyneux (1656–98) as a 'peevish old gentleman, who would not have it believed that it is possible to do better than he has done . . .'

Halley continued to correspond with Hevelius in a friendly fashion years after the trip to Danzig and had shown deep shock and remorse when an unfounded rumour of Hevelius's death reached London shortly after Halley had returned from Danzig.

The rumour of Hevelius's death came just before the news of a catastrophic fire that had burned Hevelius's observatory to the ground. The observatory was probably the most magnificent one in Europe, if not the world, and instruments, books and astronomical data were lost for ever. It is a mark of Hevelius's courage and determination that, in his late sixties, he set about building a new observatory with financial aid from benefactors from all over Europe.

In December 1680, now a Fellow of the Royal Society, Halley embarked on the 'Grand Tour' of Europe which was a cultural trip to France and Italy, traditionally considered part of a gentleman's education.

Just before he left for France, Halley saw a brilliant comet in the sky. When he arrived in Paris with his travelling companion Robert Nelson, he found that the man with whom he was staying, Jean Cassini, Director of the Paris Observatory, had also been interested in the comet. In January, Halley wrote to Hooke from Paris and said that the 'generall talk of the virtuosi here is about the Comet, which now appears, but the cloudy weather has permitted him to be but very seldom observed, whatever shall be made publick about him here, I shall take care to send you, and I hope when you shall please to write to me you will do me the favour to let me know what has been observed in England'.

In May 1681, Halley wrote to Hooke again saying that Cassini thought it possible that the recent comet might have been the same one spotted by Tycho Brahe in 1577 and that both comets had similarities to the one of 1665. It is interesting to note that Halley thought that Hooke would have difficulty in accepting the notion of the same comet appearing more than once: 'I know you will with difficulty Embrace this Notion of his, but at the same tyme tis very remarkable that 3 Cometts should soe exactly trace the same path in the Heavens and with the same degrees of velocity.'

This notion of the same comet appearing more than once must have been extremely influential when Halley was later to calculate and predict the return of the Comet that now bears his name, which he saw in 1682.

While in Paris, Halley sought out new scientific books on behalf of the Royal Society library and Robert Hooke. He also investigated and noted the size and population of Paris which he mentions to Hooke in his second letter. He found that Paris

> . . . is not soe great a Continuum of houses as London, but by reason of theire liveing many in a house it seemeth more populous, and theire bills of burialls and Christning Confirmes it, for the last yeare 1680 were buried 24411 whereas at London 20000 is reckoned a high bill, and the Christnings farr exceed ours, haveing been almost 19000, when we have ordinarily but 12 or 13000 . . .

By November 1681, Halley and Nelson had reached Rome and Halley was able to indulge his interest in the Ancient Roman remains. However his trip was suddenly cut short. He was called back by unexpected family affairs, possibly to do with his father's second marriage, and arrived in England in early 1682.

In April 1682, the year Halley's Comet returned, Halley married Mary Tooke. According to contemporary accounts she was 'a young lady equally amiable for the gracefullness of her person and the beauties of her mind . . .' and 'an agreeable young Gentlewoman; and a Person of real merit; she was his only wife, and with whom he lived very happyly, and in agreement, upwards of 55 years . . .'

Halley and his wife made their home in the village of Islington, not far from the City of London, and Halley set up a small observatory in his house. Here he began the task of observing the moon through an eighteen-year period in an effort designed to solve the problem of determining longitude at sea. Two years later he broke off his observational programme as a result of the upheaval caused by the death of his father. But it was this lengthy and arduous task that he was to resume when he became Astronomer Royal in 1720.

In the following year, 1683, Halley had two papers published in *Philosophical Transactions*, the journal of the Royal Society. The first paper was on the planet Saturn and one of its satellites. The second was on the earth's magnetism. It had long been realised that a compass needle does not point exactly due north or south but at an angle to the vertical plane through the needle's axis. This angle or magnetic 'variation' or 'declination', it was discovered, differed according to what part of the world you were in. There was a need to find a theory that could account for this magnetic variation. Halley had been interested in the ship's compass on the way to St Helena. He now proposed, in this paper, the novel but erroneous idea that the earth had four magnetic poles rather than two and that the behaviour of the needle was affected by the pole to which it was nearest. According to modern theories, Halley was wrong, but his idea was certainly ingenious and would have stimulated intellectual interest in the subject.

In 1684, Halley suffered the loss of his father who died in rather mysterious circumstances. One morning in March, Edmond Halley Snr left home and did not return in the evening nor, in fact, ever again. His wife became alarmed at his absence and offered a reward of £100, advertised in the *Gazette*, to anyone who could find him dead or alive.

Five days after he had left home, Halley's father was found by the riverside at Temple-Farm near Rochester. According to a broadside (a news sheet) of 1684, what happened was this:

> A poor Boy walking by the Water-side upon some Occasion spied the Body of a Man dead and Stript, with only his Shoes and Stockings on, upon which he presently made a discovery of it to some others, which coming to the knowledge of a Gentleman, who had read the Advertisement in the *Gazet*, he immediately came up to London, and acquainted Mrs Halley with it, withal, telling her, that what he had done, was not for the sake of the Reward, but upon Principles more Honourable and Christian, for as to the mony, he desired to make no advantage of it, but that it might be given intirely to the poor Boy; who found him and justly deserved it.

Mrs Halley sent her husband's nephew to identify the corpse. The nephew recognised the shoes from which he had previously just cut out the linings to make them more comfortable for his uncle. Whether Halley Snr was murdered or committed suicide is not known for sure, but the court verdict of the day judged that he had, in fact, been murdered.

After the death of his father, Halley found himself in a legal wrangle with his stepmother over the interpretation of the will. The man, Adams, who reported the discovery of the corpse to Mrs Halley ended up suing her for the £100 reward. So much for his 'Principles more Honourable and Christian'! The judge who conducted the case between them was Judge Jeffreys, who later became infamous for conducting the 'Bloody Assizes' in 1685, sentencing to death the Duke of Monmouth and his supporters after a failed rebellion against James II. Jeffreys awarded £20 to Adams and the rest of the reward money to the guardians of the poor boy who had discovered the corpse in the first place.

In the early months of 1684, Halley engaged himself in studying the laws of planetary motion that had been worked out at the beginning of the century by Johannes Kepler. He was especially interested in the third law which states that the square of the time taken by a planet to orbit the sun once is directly in proportion to the cube of the average distance between the planet and the sun. What particularly interested Halley was what sort of attraction the sun would have to exert on a planet for Kepler's third law to hold true. Halley had tried working out the problem and had

Kepler (1571–1630). The German astronomer and mathematician who discovered that Mars did not move in a perfect circular motion, but in an ellipse. See Chapter 15.

Newton (1642–1727). The English scientist, mathematician and astronomer. Famous for his great work known in short as the *Principia*, one of the greatest scientific books ever written. Without Halley it might never have been published.

arrived at a solution. What he really needed was geometrical proof to substantiate his answer and here he could go no further.

Robert Hooke had been working on the same problem and also claimed to have the requisite proof. Hooke seemed to be unwilling to reveal what this much desired proof was, probably because he was bluffing, even though Sir Christopher Wren, the architect and Fellow of the Royal Society, had offered a small prize (a book of the winner's choice to the value of £2) to the person who could provide the proof first. Halley, for his part, lost patience with Hooke and decided to go to Cambridge to enlist the help of Isaac Newton (1642–1727), who had a reputation as a brilliant mathematician.

In August 1684, Halley visited Newton for the first time and the results were, according to Halley's biographer Colin Ronan, '. . . so far-reaching that they can be said to have altered the whole course of physical science'.

When Halley put to Newton the problem concerning Kepler's third law, Newton recognised the problem, said he had solved it and promised to send the proof of it to Halley when he could lay his hands on it. Halley was obviously surprised and delighted at this. It also dawned on him as they spoke that Newton had made great ground in other related scientific matters that were of the first importance. In the shy and diffident Isaac Newton, Halley had found a scientific treasure trove.

On a second visit to Newton, Halley performed the greatest service to science by managing to persuade Newton to write down his ideas systematically. A lot of Newton's most important work had been formulated as early as 1665, but it seems he never had the impulse to disclose his ideas to others. Now, stimulated and exhorted by Halley, he began to write what was to be his greatest work and which would take him eighteen months to complete.

In May 1686, the year following the death of Charles II and the accession to the throne of his brother James II, the Royal Society announced publicly that they would publish Newton's work. It then came to light that the Society's finances were low and that the project might have to be shelved. Halley came to the rescue. He was determined that, in the interest of science, the book must be published as soon as possible. He therefore decided to finance the whole project himself, even though he was not rich.

Robert Hooke, who had been involved in the telescopic versus open sights controversy with Hevelius, now almost jeopardised the printing of the whole book. This happened when he made it known that he wanted to be mentioned in the preface of Newton's work, since he claimed that Newton had used some original material from him. Newton was outraged by the suggestion. Later, thoroughly sickened by the whole wrangle, Newton decided to suppress the third part of his work that included important studies on comets, planets and tides.

Halley was greatly alarmed that such important material should never see the light of day. Accordingly, he brought to bear all his considerable diplomatic skills on Newton in a bid to change his mind. Fortunately, Halley succeeded, and in July 1687 Newton's great work, the *Philosophiae Naturalis Principia*, known in short as the *Principia*, was published. Halley had referred to it as the 'divine treatise'. It is, quite possibly, the greatest scientific work ever written.

Halley's contribution to the whole project could not have been more vital: he encouraged Newton to write it in the first place; he read and corrected the proofs; he handled the row between Newton and Hooke; and, on top of that, he paid for it out of his own pocket. As Augustus de Morgan put it, writing in the mid-nineteenth century: 'But for him [Halley], in all probability the work would never have been thought of, nor when thought of written, nor when written printed.'

10

The Astonishing Edmond Halley

By early 1686, Halley had become Clerk of the Royal Society, a post that involved helping the Secretaries in matters of organising and reporting meetings and the publication of *Philosophical Transactions*.

As an official, paid employee of the Royal Society, it meant that Halley had to resign his Fellowship. Halley's first task as Clerk was to improve the output of the Society's correspondence to eminent scientists at home and abroad, on subjects including biology, ancient history, geology, geography and astronomy. The two following extracts of letters from Halley to John Wallis, his old Oxford professor, show the range of interesting material being dealt with by Halley and the Royal Society. The second extract rather amusingly indicates Halley's attitude to the French.

November 13th, 1686:

> The Royal Society has lately received an odd Inscription found lately on the basis of a pillar at Rome, if you value those things in my next I will send you a Copy thereof. There has been likewise presented to them the relation of a very strange tomb, lately found in France, supposed before Christianity in that Country. Both these are ordered for the Transactions. Likewise a relation of a little man less than a pygmie, said to have been lately presented to the French King, being 37 years old, and with a great beard, and yet but 16 inches high.

London, April 9th, 1687:

> We have had lately from France a very odd relation of an Hermaphrodite at Tholose, who being in all appearance female has yet a penis of very considerable size, but that which is the most remarkable is, that it is perforated, and is the common passage of urine, semen and the Menstrua which she has regularly. She has been hitherto habited and named as a female, but this discovery has obliged her to change the name of Marguerite to Arnold and to put on breeches. There is some difficulty to believe this story, tho it seems well attested, being from a noted physician of the place; but the bantering ridiculing humour of that light nation makes me suspect all that comes from thence.

As well as dealing with the Society's correspondence, Halley was also doing his own scientific work. In 1686, he published papers on the barometer and on the trade winds and monsoons. His paper on the winds particularly contributed to an understanding of their nature.

In 1688, the year James II fled and William of Orange was invited to England, Halley became the proud father of two daughters. Although they were born in the same year, they were not twins. Little is known of them except that they were called Margaret and Catherine and were alive after their father's death in 1742.

During this period, and for the next eight years, Halley's scientific work continued with the publication of several papers on such diverse subjects as the evaporation of sea water by the sun, magnetism, the distance of the sun from the earth, the time and place of Julius Caesar's landing in Britain and a criticism of Pliny's *Historia Naturalis*.

Halley was interested in an extraordinary variety of things. Examples of the topics he was dealing with at meetings of the Royal Society are shown in the extracts taken from the contents of the *Journal Book* of the Royal Society for the year 1689:

> July 17th, 1689:
> Halley said that in blowing up houses, or any other great explosion of Gun-powder the windows near adjoyning are not blown inwards into the houses, as is generally supposed, but alwaies outward into the street, of which he assigned the reason that the air being rarified and consumed by the flame, the pressure of that in the rooms of houses exceeding that in the street, the windows come to be thrust out thereby.

> July 24th, 1689:
> Halley said, that before the late great frost, he having used to water Rosemary, and other tender plants with soap sudds much diluted, and only a little blewish with soap, he found that these flowers throve well, and bore the hard winter better than in other gardens. This was thought to be a very proper meame [measure] for the earth by reason of the oyle and alchaly-salt in soape, but what degree of salt was proper in this case was a question not easy to determine.

> November 27th, 1689:
> Halley said the way that Poultrers knew, whether wild-fowl is fresh, or no, is by the foot, for they conclude, that the foot being dry it is certain, they are stale, if not, they are esteemed fresh.

> January 22nd, 1689/90:
> Halley read a Discourse tending to prove, that Julius Caesar first landed in Britain Anno 55 ante Christum on August 26th in the afternoon to the Northward of the south forland and in all probability in the Downs near the place, where now the Town of Deale stands: this he made out by the description given of this Expedition in Caesar's Commentarys, and by Dion Cassius in his 39th. book.

> January 29th, 1689/90:
> Halley related some thing like this, to have befallen himself, in takeing a large Dose of Theriaca Andromachi, wherein he believed the Opium not to have been well mixed; for instead of sleep, which he did design to procure by it, he lay waking all night, not as if disquiet with any thoughts but in a state of indolence, and perfectly at ease, in whatsoever posture he lay.
>
> February 5th, 1689/90:
> Halley related, that in the Barbadoes they had formerly made their Hedges or Fences with the prickle-pair or Indian Figg, which by their thickness, and long prickles were not to be attempted by Hoggs or any sort of cattle.

Something of Halley's character, in particular his tact and deference, can be seen in an extract of a letter to William Molyneux in which Halley apologises for the delay in writing:

> I assure you, Sr. it is in no measure to be attributed to a want of respect, for your person, whom all that knows, must needs acknowledg singularly endowed with a talent for reasoning aright, and judging truly in all things brought before you; but so it is, that a certain sluggish indisposition of mind, has been the frequent occasion of my committing several indecencys of the like kind, and in your particular, with shame I must own it, this is not the first time; there is nothing but your good nature, can set me right in your esteem which I promise myself will be done, when by future diligence, I shall make it appear that I have cured myself of that torpid maladie, wherby I have so well deserved to loose many friends.

By 1689 Halley had become interested in deep-sea diving. In an unpublished paper of 1689, he describes the problem of breathing faced by divers and suggests improvements for the diving bell that had existed in some form or other since the mid-sixteenth century.

The diving bell was basically a large, bell-shaped container which was lowered down into the sea. From it, divers could breathe in fresh air (trapped inside the bell), rather than having to swim right up to the surface of the sea.

Halley's improved diving bell was cone-shaped with a flattened top three feet across and had a window of plate glass. It was five feet high and five feet across the bottom, which was open. Inside, the bell was fitted with a bench for men to sit on.

One of the problems that beset the bell was that, as it was lowered into the sea from a ship, water entered it from the bottom because the air inside the bell contracted more and more as the bell got deeper. The problem was to get rid of this unwanted water and Halley thought of a way of doing so by using two large barrels, filled with compressed air. Each one was cased in lead and had a bung-hole in the bottom and a valve in the top.

These barrels were then lowered alternately from a boat into the water and down next to the bell. Each one was then in turn brought under and into the bell.

The air was then released from the barrel through the valve at the top. As the air was released, the bung-hole in the bottom of the barrel was simultaneously opened and the bell's water rushed into the barrel, replacing the compressed air. When the water had filled the barrel, the latter was hauled up and the other barrel was lowered down for the process to be repeated.

Halley himself was brave enough to try out his version of the diving bell and in an unprinted paper he describes the painful pressure he felt in the ears when descending into the water:

> When we lett down this engine into the sea we all of us found at first a forceable and painful pressure on our Ears which grew worse and worse till something in the ear gave way to the Air to enter, which gave present ease, and at length we found that Oyle of Sweet Almonds in the Ears, facilitated much this admittance of the Air and took the aforesd pain almost wholly.

Halley also recorded that he was able to keep three men under the water at a depth of ten fathoms for one and three-quarter hours.

As well as the improvements Halley made to the diving bell, he also designed a diving suit and a helmet which was connected to the bell by two long flexible pipes. These had the job of feeding fresh air to the diver and taking away the air he breathed out.

Halley's confidence in his excursion into the underwater world is shown by the fact that he set up a salvage company whose share prices were quoted in the trade journal of the day.

In 1691, Edward Bernard resigned as Savilian Professor of Astronomy at Oxford, the professorship founded by Sir Henry Savile (1549–1622). Halley was one of the applicants for the vacant post.

It was statutory, at that time, that the successful candidate for the post should, as a member of a teaching staff, be an orthodox member of the Church. Halley's problem was that there was an unsubstantiated rumour that he did not believe in God, which, if it were thought to be true, would prevent him from being accepted for the post. There is some doubt as to the exact nature of the accusations levelled at Halley. William Whiston, a mathematician, claimed that Halley was not faithful to the Church of England. Against this it must be remembered that Whiston was writing more than fifty years after this time. Also one must bear in mind the fact that Whiston's own Arian beliefs prevented him from being elected to the Royal Society, and this may have coloured his account.

Halley himself referred to the election in a letter to Abraham Hill, in which he says:

> This business [i.e. diving] requiring my assistance, when an affair of great consequence to myself calls me to London, viz. looking after the Astronomy-Professor's place at Oxford, I humbly beg of you to intercede for me with the archbishop Dr Tillotson, to defer the election for some short time, 'till I have done here, if it be but a fortnight: but it must be done with expedition, lest it be too late to speak. This time will give me an

Savile (1549–1622). The English scholar who founded the Savilian Professorship of Astronomy and Geometry at Oxford, and which Halley was so keen to have awarded to him.

> opportunity to clear myself in another matter, there being a caveat entered against me, till I can shew that I am not guilty of asserting the eternity of the world.

In the event, Halley was summoned before Bishop Stillingfleet, who had acted as chaplain to Charles II. He was the most famous bishop of his day. Apparently, the bishop was not satisfied with Halley's answers to his questions because Halley had to undergo a second bout of questioning from Stillingfleet's chaplain, Richard Bentley. Bentley was a harsh, proud and avaricious man and it is easy to imagine that the relatively liberal attitude and easy manner of Halley would have rankled with the clergyman. There is no doubt that Halley did believe in God. Also, if he did have any unorthodox views, it is most likely that he would have kept them to himself, since he did not publish a paper on the biblical Flood for thirty years for fear that the Church might take it amiss.

As it turned out, Halley failed to get the Savilian professorship. This was in spite of recommendations both from his old college, Queen's, Oxford, and the Royal Society. The opposition to Halley must have been considerable since he had excellent credentials for the post and the decisive factor may well have been charges levelled at him by his colleague, the Astronomer Royal, John Flamsteed.

Flamsteed's hostility towards Halley got the better of him when he heard that Halley was applying for the post. Flamsteed immediately wrote to Isaac Newton and urged him to oppose Halley's application on the grounds that, if he were elected, he would corrupt his pupils with 'lewd discourse'. He also claimed that Halley had infidel friends. Coming from the Astronomer Royal, these were serious accusations.

Flamsteed's antagonism to Halley goes as far back as 1686. Flamsteed had then been defensively hostile to Halley because Halley had criticised his tide tables. Flamsteed's tables were based on the tides at London; but he also applied them, incorrectly, to other ports including Dublin, whose tides Halley himself had investigated with information from his friend William Molyneux, who lived in Dublin. Halley could see from his own work that Flamsteed's tables were in error.

Flamsteed, who was a moody and dour man and who had spent a long time and much energy on his tables, was furious with Halley and from then on began to show his resentment in letters to Newton. He even went so far as to cast aspersions on the quality of Halley's catalogue of the southern stars and to accuse Halley of plagiarising ideas on the earth's magnetism from a Mr Perkins, a mathematics master at Christ's Hospital.

When Halley published an improved theory of the earth's magnetism in 1692, Flamsteed again tried to detract from Halley's achievement. Even though the theory is, according to modern ideas, no longer tenable, it was nevertheless at the time a considerable effort at coming to terms with the phenomenon.

Another source of Flamsteed's hostility towards Halley was the latter's friendship with Robert Hooke. Hooke, it seems, was liable to have a joke at Flamsteed's expense but one can imagine that Flamsteed must have been a tempting target for teasing.

It is a great pity that Flamsteed, who was himself a very able and much respected astronomer, should have felt such hostility towards Halley. Halley never seems to have been hostile in return, although he did vindicate himself, in 1692, of Flamsteed's criticism of his observations of the southern stars made at St Helena.

Since the earliest times, comets have been seen in the skies and their unpredictable and often spectacular appearances could remain visible for weeks. It is hardly surprising that when men relied upon the reassuring stability of the seemingly fixed patterns of glittering stars, comets were feared as harbingers of misfortune or, at best, indicators of a change in the status quo.

To ancient astronomers, the nature of comets was mysterious. Aristotle, for example, thought that comets were not celestial bodies that moved in orbit round the sun but the result of the upper part of the earth's atmosphere, which he thought was very hot and dry, suddenly igniting. Thus, he concluded that a comet was always closer to the earth than was the moon.

After it had been established in the late sixteenth century that the earth went round the sun and not vice versa, it was wondered, contrary to what Aristotle thought, whether a comet could exist beyond the moon. The issue was settled by the evidence provided by the Danish astronomer Tycho Brahe of the bright comet of 1577. From his data, Tycho could now prove that comets lay further from us than the moon.

A comet was now recognised as a celestial body. But the nature of its motion through the sky was still a problem. Matters were helped by the publication of Newton's *Principia* in which it was stated that the sun must have a gravitational influence on a comet. On this basis it was calculated that three types of cometary path were possible. A comet's orbit could be parabolic (U-shaped), elliptical (elongated-circular shaped) or hyperbolic (U-shaped but with the two arms diverging and not parallel).

In 1695, with these possible cometary paths in mind, Halley began to examine all the observational information he could find up to the end of the seventeenth century, in an effort to find out about the nature of cometary paths. He then calculated the paths of twenty-four different comets. The mathematics involved were laborious and difficult since, apart from anything else, he had to take into account the effects on the comets from the gravitational fields of the outer planets.

As his work on the comets progressed, Halley became more and more sure that cometary paths were elliptical or elongated circular, which would mean that comets had long orbits around the sun. He was also sure that the Comet of 1682 and the comet of 1531 had orbital characteristics very similar to the Comet of 1607. These comets he would, in due course, correctly reckon to be the same one. In letters to Newton at this time, we can see Halley getting closer and closer to his successfully predicting the return of the Comet that now bears his name.

In a letter of September 28th, 1695, Halley asked Newton

> to procure for me of Mr Flamsteed what he has observed of the Comett of 1682 particularly in the month of September, for I am more and more confirmed that we have seen that Comett now three times, since ye Yeare 1531, he will not deny it you, though I know he will me.

The exact date of the next letter to Newton is not known, but it must have been written before 15th and after 1st October 1695:

> Honoured Sr. In answer to yours of the 1st of October, I give you many thanks for yr Communication of the Observations of the Comet of 1682 which next after that of 1664 I will examine, and leave it to your consideration, if it were not the same with that of 1607, and when your more important business is over, I must entreat you to consider how far a Comets motion may be disturbed by the Centres of Saturn and Jupiter, particularly in its ascent from the Sun, and what difference they may cause in the time of the Revolution of a Comet in its so very Elliptick Orb.

The next letter to Newton is dated October 15th, 1695. In it Halley tells Newton that he will 'examine the Comet of 1682 and send you the result, hoping it will give me no great trouble, because the observations I presume are exact . . .'

Halley tells Newton in a letter of October 21st, 1695 that he has 'almost finished the Comet of 1682 and ye next you shall know whether that of 1607 were not the same, which I see more and more reason to suspect'.

Halley's calculations for the comet of 1680 predicted that its return would be in 575 years time, a figure now considered to be incorrect. But for the comet of 1682, Halley predicted that it would return in 1758. He realised that it was unlikely that he would still be alive in that year (he would have been 102) and so he expressed the hope that astronomers would scan the skies for the comet at the predicted time.

Halley's results were later published in 1705 as *A Synopsis of the Astronomy of Comets* and also in *Philosophical Transactions*. But he had made his findings known to the Royal Society in June 1696, as is recorded in the *Journal Book* of the Society:

> June 3rd, 1696. Halley produced the Elements of the Calculation of the Motion of the two Comets that appear'd in the Years 1607 and 1682, which are in all respects alike, as to the place of their Nodes and Perihelia, their Inclinations to the plain of the Ecliptick and their distances from the Sun; whence he concluded it was highly probable not to say demonstrative, that these were but one and the same Comet, having a Period of about 75 years; and that it moves in an Elliptick Orb about the Sun, being when at its greatest distance, about 35 times as far off as the Sun from the Earth.

The Comet was spotted on Christmas Day 1758 by a Saxon amateur astronomer called Palitzsch, sixteen years after Halley's death. It was duly named after the man who, anticipating the success of his own prediction, had asked that it might be acknowledged that the predicted return 'was first discovered by an Englishman'.

Thanks to Halley, comets were now shown to be subject to laws like other natural phenomena and Newton's theory of universal gravitation was, for the first time, used by an astronomer with practical success.

In December 1694, the year Queen Mary died, Halley read a paper to the Royal Society on the cause of the biblical Flood. The paper was not published until thirty years later since, at this time, Halley was worried that the Church might see

Palitzsch, the Saxon farmer with a serious interest in astronomy which brought him renown when he was the first to sight, on Christmas Day 1758, the return of the Comet predicted by Halley.

him as a scientist meddling with the sacred scriptures. This might confirm them in their suspicions of Halley's alleged unorthodoxy that had prevented him from getting the position of Savilian Professor at Oxford.

Halley thought that the account of the Flood in Genesis had enough detail to suggest that it was based on a much fuller record and yet not so obviously accurate that it seemed to be a result of a divine revelation. To suggest that the biblical account was not a revelation from God but an imperfect version of an earlier account was to verge on heresy at that time.

Halley studied the story of the Flood and found the description of the forty days of rain covering the earth, and the animals in the ark, scientifically implausible. He thought it was most probable that the Flood was caused by a comet coming too near the surface of the earth and causing a gravitational upheaval. This, he thought, would account for the Flood suddenly happening out of the blue and, logically, it is sound. However, what Halley failed to realise was that a comet's mass is not sufficient to cause such a great disturbance.

In the mid 1680s, the English government realised that debasement of the currency by the illegal practice of clipping bits off silver coins and re-smelting them was becoming a serious problem. The only answer seemed to be to keep in circulation coins with milled edges and to withdraw unmilled coins. So, a massive recoinage programme was started.

Accordingly, a number of mints were set up and Sir Isaac Newton was put in charge of the project by the Chancellor of the Exchequer, Charles Montagu, in 1696. In the same year, Newton appointed Halley as Deputy Controller of the Mint at Chester.

Chester was not an ideal place for Halley to continue his astronomical work and he was glad when his two years there came to an end. However, it is typical of Halley that he should make the most of his stay. He became interested in the town's ancient history, especially a Roman altar piece made from local stone; he carefully observed the tidal River Dee; he reported the occurrence of a freak hailstorm that killed hens, lambs and a dog; and, during an excursion to north Wales, he tested out his theory of gauging heights by barometric readings on Mt Snowdon.

In a letter dated October 12th, 1696, seventeen days before his fortieth birthday, Halley wrote to Sir Hans Sloane (1660–1735), a Fellow of the Royal Society, and told him of the matters of interest that he had found in Chester:

> There are severall Antiquities in and about this place in custody of private persons, of which I will take care to give you a description; and the situation of the City is very remarkable; being at the place where the river Dee ceases to flow, upon any other than spring Tides. And the walls and all publique buildings are of a stone which is afforded by Quarries which are upon the spott, and in many places appear in the Ditch of the Town, which convenience I suppose occasioned the Romans to found the City here, which is square like the Roman Castra, and each side is about 600 or 700 Yards. and here was for a long time the head Quarters of the legio XX. called Victrix.

Halley's circumstances were made difficult at Chester by trouble at the mint. Two of the clerks were, it seems, involved in some racket with the connivance of the Master of the Mint and they could see that Halley and the Warden of the Mint, Mr Weddell, were not going to stand for any corrupt practices. They therefore conspired to harass and make false accusations against Halley and Weddell.

In a letter dated October 25th, 1697, Halley tells Sloane: '. . . In the mean time my heart is with you, and I long to be delivered from the uneasiness I suffer here by ill company in my business, which at best is drudgery, but as we are in perpetuall feuds is intollerable.'

Matters reached the point that Halley was forced to write to Newton, explaining the situation and stating that he would be ready to face Lewis, the principal instigator of the trouble, before a tribunal. Halley wanted Lewis to answer various charges including the throwing of an inkstand at Weddell.

As it happens, Newton already knew of Halley's dissatisfaction with his position at the mint and offered him two other situations of employment. Halley turned these down, presumably because he found them unsuitable, despite his desire to leave Chester as soon as possible. The mint was finally closed in 1698 and Halley was able to return to London.

In that same year, 1698, Halley spent some time with Czar Peter the Great (1672–1725), Emperor of Russia. Peter had been invited by William III to come to England to study methods of shipbuilding in order to strengthen his navy against the Turks in the Black Sea.

Peter, who was then twenty-six years old, stayed at Sayes Court in Deptford, the home of John Evelyn on lease at the time to Admiral Benbow. The Czar was a dynamic and energetic young man with a lively sense of humour. He could also have violent fits of temper and be extremely cruel. He and his circle enjoyed an extravagant and often riotous lifestyle – so much so that extensive damage was done to the house and gardens of Sayes Court during their stay there. Furniture, pictures and up to 3,000 windowpanes were destroyed.

Halley was given the task of entertaining the Czar and informing him of all the latest scientific news and developments. Peter enjoyed both Halley's considerable knowledge and his unpompous, friendly manner. There is even a rather dubious story that, on one occasion, the Czar was given a ride round the garden in a wheelbarrow pushed by Halley. The ride came to an unceremonious end with the passenger climbing out of a prized hedge!

After the episode at the Chester mint, Halley's next major undertaking was a voyage on the Atlantic Ocean. The object of the voyage was primarily to record the variation of the compass as they travelled from place to place and to determine the longitude and latitude of ports the ship called at. The instructions he received also required him to visit the English West Indies on the return home. It was extremely rare for a landsman to be made captain of a ship of the Royal Navy and a mark of the esteem in which Halley was held that he was appointed to this post.

Preparations for the voyage had, in fact, begun as early as 1693, but various delays, including Halley's post at Chester, had postponed it till 1698. By August 1698, the ship, called the *Paramour*, a three-masted vessel of the type known as a

Sir Hans Sloane (1660–1753). Contemporary and friend of Edmond Halley whose benevolence and influence caused many London landmarks to be named after him.

pink, was ready with a crew of twenty men and enough food for a year. The voyage was under the patronage of William III and it was the first scientific voyage to receive a royal commission, long before the time of Captain Cook. As Captain S. P. Oliver wrote in the late nineteenth century: 'We do not think of him as a sailor; and yet, previous to Cook, Capt. E. Halley was our first scientific voyager.'

In late November 1698, the *Paramour* set sail from England and after a fortnight they reached Madeira. Here they parted company with Admiral Benbow and his ships, who had been safeguarding their journey since the Isle of Wight. From Madeira, Halley wrote to Josiah Burchett, Secretary to the Admiralty, on December 19th:

> Honoured Sr. On the sixteenth Instant I arrived at this Island together with the *Glocester*, the *Falmouth*, the *Dunkirk* and *Lynn* frigots, under the Command of Rear Admirall Benbow. By reason of the Holydays it was not possible for the Shipps to have their Wines on board before this day, wch occasioned the Admirall to leave the Island the same night he arrived, being unwilling to waite so long. I have gotten my self dispatcht, and shall persue my Voiage with the first wind it being now Calm. I thought I ought to give their Lopps [Halley's word for 'Lordships'] an account of our arrivall here, not finding that there were any letters left for you by the Admirall here; who left the Island in all diligence. I am Your Honours most Obedt. Servant Edm. Halley.

From Madeira, the *Paramour* headed for the Cape Verde Islands (off the west coast of Africa) via the Canaries. Approaching their destination, they were fired on by an English ship which mistook them for pirates.

Halley relates the incident to Burchett in a letter dated April 4th, 1699:

> . . . instead of saluting us, to fire at us severall both great and small shott. We were surprised at it, and beliving them to be pirates, I went in to windward of them and bracing our head sailes to the Mast, sent my boat to learn the reason of their firing. They answered that they apprehended we were a pirate, and that they had on board them two Masters of vessells, that had been lately taken by pirates, one of which swore that ours was the very shipp that took him; whereupon they thought themselves obliged to do what they did in their own defence.

It was in the early months of 1699 that Halley began to have trouble with the crew. On one occasion, the boatswain deliberately steered a course contrary to Halley's orders and Halley had to correct what the boatswain claimed was a 'mistake' himself.

Their next stop was on the west coast of Brazil, where they were greeted by the Portuguese governor who invited them ashore to take on supplies of water. But the friendliness of the governor may have been a façade for, as Halley writes in the same letter to Burchett:

> . . . the Portuguez, as farr as I could guess, were willing to find pretences to seize us, and tempted us severall times to meddle with a sort of wood they call Poo de Brasile which is an excellent dye, but prohibited to all forreigners under pain of confiscation of Shipp and goods. I being aware of their design absolutely refused all commerce with them, and having gotten our water we arrived here [Barbados] in three weeks, the second of this month: Our whole shipps company is here in perfect health and our provision proves very good.

Halley left Brazil and, disturbed by the uneasiness of the crew, decided to make for Barbados with a view to taking on new officers there. By March, as they were reaching Barbados, Halley had a serious problem with one of his officers whose unmanageable behaviour eventually forced Halley to arrest him and sail home to England, cutting the voyage short.

Halley related what happened with this officer, Lieutenant Harrison, to Burchett in a letter sent after Halley had returned to England in June 1699:

> But a further motive to hasten my return was the unreasonable carriage of my Mate and Lieutenant, who, because perhaps I have not the whole Sea Dictionary so perfect as he, has for a long time made it his business to represent me, to the whole Shipps company, as a person wholy unqualified for the command their Lopps have given me, and declaring that he was sent on board here, because their Lopps knew my insufficiency . . . On the fifth of this month he was pleased so grosly to affront me, as to tell me before my Officers and Seamen on Deck, and afterwards owned it under his hand, that I was not only uncapable to take charge of the Pink, but even of a Longboat; upon which I desired him to keep his Cabbin for that night, and for the future I would take charge of the Shipp my self, to shew him his mistake: and accordingly I have watcht in his steed ever since, and brought the Shipp well home from near the banks of Newfound Land, without the least assistance from him.

In July 1699, Lieutenant Harrison was court-martialled and it was revealed that three years previously he had written a small book on the problems of determining longitude at sea. Halley had been one of those who had found the ideas in the book wanting. Obviously Harrison had resented Halley's criticism and had decided to get his revenge on board the *Paramour*.

In a letter to Burchett dated July 4th, 1699, Halley tells what happened at the court-martial and comments on the lightness of the sentence passed on Harrison and the other officers:

> Yesterday at the Court Martiall I fully proved all that I had complained of against my Lieutenant and Officers, but the Court insisting upon my proof of actuall disobedience to command, which I had not charged them with, but only with abusive language and disrespect, they were pleased only to reprimand them, and in their report have very tenderly styled the abuses I

Peter the Great (1672–1728). The Russian Czar who visited London in 1698 and to whom Halley was assigned as a companion.

> suffered from them, to have been only some grumblings such as usually happen on board Small shipps.

Undeterred by his experiences of the first voyage, in September 1699 Halley set sail on another voyage with the same scientific objectives. Despite the fact that the first voyage had ended abruptly, Halley had gathered valuable information concerning the variation of the compass and the latitudes and longitudes of ports he had stopped off at. It was hoped now that the second voyage would bear even more fruit.

Halley headed for Madeira first of all, but then fearing 'Sally Rovers' – Moroccan pirate ships – he changed course for the Canaries where they ran into stormy weather.

A tragic accident then occurred when the ship's boy, Manley White, fell overboard. Despite efforts to save him, the waves were too high and the boy was drowned. The incident had a profound effect on Halley who, it is said, could never refer to it subsequently without tears in his eyes.

After stopping at the Cape Verde islands off the west coast of Africa, Halley set the *Paramour* on a course for South America. He was, all this time, constantly making observations of the magnetic variation and longitude and latitude.

They reached Rio de Janeiro early in December 1699 and by the end of the month had set off southwards. Soon they encountered stormy, freezing weather and high seas. Somewhere north-east of the Falklands, Halley recorded spotting diving birds with long swan-like necks, and also large, finned sea creatures.

Further south, now in the month of February with the temperature reaching freezing point, the *Paramour* came upon three islands that were not marked on the map. Halley describes these islands in a letter to Josiah Burchett dated March 30th, 1700, about two months after the event:

> . . . we fell in with great Islands of Ice, of soe Incredible a hight and Magnitude, that I scarce dare write my thoughts of it, at first we took it for land with chaulky clifts, and the topp all covered with snow, but we soon found our mistake by standing in with it, and that it was nothing but Ice, though it could not be less than 200 foot high, and one Island at least 5 mile in front, we could not get ground in 140 fadtham.

These mystery islands must have been giant icebergs. The next day they found themselves in danger from the icebergs and the fog, as Halley tells Burchett in the same letter:

> . . . we were in Imminent Danger to looss our ship and lives, being Invironed wth Ice on all Sides in a fogg soe thick, that we could not see it till was ready to strike against it, and had it blowne hard it had scarce been possible to escape it.

Halley and his crew escaped to tell the tale and took a northward course into more clement weather.

The *Paramour* then went to St Helena, the island where Halley had charted the stars of the southern hemisphere, to take on fresh water and provisions. As Halley tells Burchett: '. . . went to St Helena, where the continued rains made the water soe thick with a brackish mudd, that when settled it was scarce fitt to be drunke; all other necesarys that Island furnishes abundantley.'

From St Helena, they made two more stops before steering for Pernambuco, a Brazilian province. Here a self-styled English consul, called Hardwicke, had Halley arrested on suspicion of being a pirate. While Halley was held, the ship was searched but, naturally enough, nothing incriminating was found. Halley was released and Hardwicke blamed the Portuguese, who controlled Recife, the capital of the province, for the apparent mistake.

In early May, the *Paramour* set off for Barbados and, when they arrived there, they found the island struck with disease, as Halley writes to Burchett:

> . . . I found the Island afflicted with a Severe pestilentiall dissease, which scarce spares any one and had it been as mortall as common would in great measure have Depeopled the Island. I staied theire but three days, yet my selfe and many of my men were seazed with it, and tho it used me gently and I was soon up again yet it cost me my skin, my ships company by the extraordinary care of my Doctor all did well of it, and at present we are a very healthy ship . . .

They arrived at Bermuda on June 20th, 1700. The *Paramour* was refitted for the trip home. They then reached Newfoundland where they were almost shipwrecked in a thick fog and then fired upon by New England fishermen who thought they were pirates. After that, however, they made their way homewards and, without incident, finally arrived at Plymouth Sound on August 27th, 1700.

The voyage had been a great success. Despite danger from icebergs, storms, disease and being fired on, Halley had brought the ship safely home with the loss of only one life.

The scientific results of the voyage were published in 1701 in the first edition of Halley's Atlantic Chart, dedicated to King William III who had been the patron of the expedition.

On the Chart, for the first time, lines were shown to indicate the distribution of the earth's magnetism. These lines were curved and went through those places that had the same variation of the compass. Halley called them 'curve lines'; they were often, subsequently, called 'Halleyan lines' and are now known as 'isogonic lines'. It is thought that these lines were the first of their kind to be printed and they have been used ever since on maps of a similar type.

In 1702, the year King William III died, Halley changed and improved his Atlantic Chart and it was republished as the World Chart in that year. The map proved to be extremely popular and ran into many editions – a revised edition being printed in 1758, sixteen years after Halley's death.

Before Halley's naval career ended, he had one last assignment: he was commanded by King William III, in 1701, to chart the tides and coast of the English Channel. It is quite possible that, since political relations with France were deteriorating at that time, Halley was told to pick up any information he could on the French navy.

The War of the Spanish Succession broke out in 1702. It arose after the King of Spain, Charles II, died without direct heir, leaving the throne to a grandson of the French king, Louis XIV. However, under two treaties of 1698 and 1700, Louis XIV had promised that no member of his family would press for the Spanish throne. King William III realised that France united with Spain would be a formidable power and so he formed the Grand Alliance with Austria, Holland and Prussia. War was then declared on France in May 1702.

Another of Halley's talents was his skill in engineering. Since British warships were to be used in the Adriatic Sea, Halley was sent by Queen Anne (who had succeeded King William III in 1702) to advise the Emperor Leopold I, in Vienna, on how to fortify the ports of Trieste and Bocari.

Travelling through Holland and Germany, Halley arrived in Vienna where he met the English ambassador, George Stepney. Straightaway, Halley went to Istria and, despite obstructive behaviour from the Dutch, he surveyed the harbour fortifications of Trieste and Bocari. He then returned to Vienna. The Emperor was impressed by Halley's intelligent recommendations and gave him a diamond ring from his own finger. He also gave him a written commendation to take back with him to Queen Anne.

Halley returned to England but very shortly set off again to Vienna to supervise the necessary construction work to be done on the harbours. On his journey there, he stayed at Hanover and there dined with the Queen of Prussia and the future George I of England.

From Vienna, Halley was again sent to Istria where he was joined by the Emperor's chief engineer. They found Trieste in need of repairs but Bocari seemed to be adequately fortified. Halley finally returned to England in November 1703.

In October 1703, the post for the Savilian professorship at Oxford again became vacant after the death of the Reverend John Wallis. The situation had changed since Halley's last and unsuccessful attempt for this position: Bishop Stillingfleet, who had headed the religious opposition to Halley, was now dead; meanwhile, Halley had greatly increased his influence and reputation. Only Flamsteed remained hostile, writing in a letter at that time that Halley was swearing and drinking like a sea captain.

In spite of Flamsteed, Halley got the appointment and was Savilian Professor of Geometry for the rest of his life. His new residence in Oxford was in New College Lane, and can still be seen today with a commemorative plaque on it. The house had been divided into two by Wallis's son and so Halley had one half of the original building. On top of his house, Halley had a small observatory built. It seems he did not move into residence there until 1705 and after 1713 he was living permanently in London again. He still kept a room, though, in his house in Oxford for occasional use when he visited there.

One of Halley's immediate tasks on becoming Savilian Professor was to translate and edit, with David Gregory, the work of the Greek mathematician Apollonios.

Apollonios lived in Alexandria at the end of the third century BC. He had specialised in the study of curved lines in geometry.

Halley first tackled a minor work of Apollonios, a small part of which had already been translated into Latin by Edward Bernard and revised by David Gregory. The problem was that the text had only survived in Arabic, and not Greek, and Halley had no knowledge of Arabic. Undeterred by this, he used the Latin translation as a key to the Arabic and was able to translate the rest of the text. It is said that Dr Sykes, the great orientalist of his day, was amazed at what Halley had been able to achieve.

Halley then proceeded to work on Apollonios's great work, the *Conics*, with David Gregory. The work was in eight sections but the last section of the text was missing. Before finishing his quota of the translation work, Gregory died and Halley had to do most of the work himself. The most difficult part was reconstructing the missing eighth section. Halley set about the task, proceeding from his knowledge of the subject and of the style of Apollonios's writing from the other seven sections. Eventually, with the help of comments written in Latin on the eighth section by the mathematician Pappus (who had seen it), Halley was able to reconstruct the missing work. It was a brilliant achievement and for this and other translations, such as his edition of the *Spherics* of Menelaus, he was given the degree of Doctor of Civil Law by Oxford University in 1710.

Halley became involved in another difficult situation which arose after 1705 between the Royal Society and Halley's antagonist and colleague, John Flamsteed.

The problem arose over Flamsteed's refusal to publish his astronomical observations, even though, as Astronomer Royal, he was technically a public servant. Flamsteed took the view that he had had to pay for his own instruments, he received a paltry salary from the Crown and had to supplement it by teaching privately; and he felt that his observational results were his own property, to be published at his own discretion.

Sir Isaac Newton, by then President of the Royal Society, needed Flamsteed's results for revising his own lunar theory. He therefore put pressure on Flamsteed to hand them over and eventually succeeded. Flamsteed reluctantly sent his star catalogue to the Royal Society who set up a committee (which included Halley) to examine it.

Flamsteed was then asked to complete and edit his work so that it could be printed – the cost for which would be met by Queen Anne's consort, Prince George of Denmark. However, Prince George died in 1708 and the printing that had been proceeding for two years, despite the unhelpful behaviour of Flamsteed, ground to a halt.

The Royal Society, at this point, lost patience with the whole affair, especially in the light of Flamsteed's begrudging attitude. As a result, Queen Anne was persuaded to issue a royal warrant in which extra powers were granted to the President of the Royal Society. These included the right to obtain from the Astronomer Royal a decent copy of his annual observations by at least six months after the end of the year.

In 1711, the printing of Flamsteed's catalogue proceeded once more. But, since Flamsteed had been so unco-operative in putting his work into fit condition for the printer, the Royal Society decided to entrust Halley with the task of finishing and editing it.

Halley set about his task and it seems from a letter written to Flamsteed that he did his best to make a good job of it and to collaborate with Flamsteed:

> Though I am credibly informed that these sheets have been, from time to time, sent you from the press, yet, lest it should be otherwise, I have now sent you the catalogue of the fixed stars intended to be prefixed to your book; having spared no pains to make it as complete and correct as I could, by help of the Observations you have given us, made before the year 1706. I desire you to find all the real faults you can, not as believing there are none, but being willing to have a work of this kind as perfect as possible: and if you signify what's amiss, the errors shall be noted, or the sheet reprinted, if the case require it. Pray govern your passion, and when you have seen and considered what I have done for you, you may perhaps think I deserve at your hands a much better treatment than you for a long time have been pleased to bestow on Your quondam friend, and not yet your profligate enemy (as you call me), Edm. Halley.

Halley's edition of Flamsteed's catalogue appeared in 1712 as *Historia Coelestis*. Absolutely furious with what he saw as gross liberties taken by Halley with his work, Flamsteed was goaded into publishing another edition worked on completely by himself. He also tried to suppress the *Historia Coelestis* and in 1715 managed to burn, with the exception of one volume, three-quarters of the total number of copies issued, making 'a sacrifice', as he put it, 'to heavenly truth'. His own edition – *Historia Coelestis Britannica* – was published posthumously and is a testimony to the skill and the labour of this talented but difficult man.

In 1719, the Astronomer Royal, John Flamsteed, died. Halley's astronomical career culminated in his being elected the second Astronomer Royal in 1720.

Up to 1720, Halley had published several important papers on subjects including meteors and magnetic variation in 1714 (the year Queen Anne died); novae (stars that suddenly increase their brightness) and nebulae (clouds of dust and gases in space) in 1715; aurorae (phenomena of glowing light seen near the earth's poles); and in 1716 ways of calculating the distance of the sun from the earth using the passages or transits of Venus across the sun's disc.

In 1717, Halley published a paper on the fixed stars. Halley had examined, as far back as 1710, the positions of the stars recorded by Ptolemy in the second century AD and had found that they were significantly different from the positions that he and others at that time had recorded. After some calculations, Halley concluded that the discrepancy between Ptolemy's star positions and those of Halley's time was due to the fact that the stars must have moved, and not that Ptolemy's instruments were inefficient.

The conclusion that the stars must have moved was revolutionary, because up till then stars were thought of as being fixed and stationary. Indeed, to the casual observer, the constellations do seem to create the same patterns in the sky year after year and it is easy to assume that they do not move. What Halley realised was that it is because the stars are so distant from the earth that it would take some 1,500 years to notice any significant change in their positions. It was not until the early nineteenth century, when instruments had been improved, that observational proof verified Halley's theory.

As Astronomer Royal, Halley went to his new work at Greenwich. There, to his dismay, he found that all Flamsteed's instruments had been removed by his widow, who had, in fact, a perfect legal right to do so. With the help of a royal grant of £500, Halley was able to refurbish the observatory. His first and main task, and one which would take him over eighteen years to complete, was to find a way of determining longitude at sea. In 1722, he began this marathon of a task.

Halley reckoned that the answer to the problem of determining longitude at sea lay in observing the moon's positions over an eighteen-year 'sarotic' period, after which the moon's irregularities are again more or less repeated over another eighteen-year cycle. With the moon's motion known, Halley thought it could be used as a giant clock hand indicating a standard reference time against the backdrop of stars, with which the mariner could compare the ship's local time and so be able to work out his longitude.

It was this eighteen-year observational programme that Halley had begun in 1683 but which he had abandoned after the death of his father in 1684. He now became so preoccupied with it that, in 1727, he was uncharacteristically dilatory in presenting a fair copy of his annual observations for the benefit of others. As Astronomer Royal he was obliged to do this, and Sir Isaac Newton, the President of the Royal Society, had to remind him of his duty. Perhaps it was the substantial cash reward offered by the government for the solution of the longitude problem that momentarily distracted Halley from his other duties.

The solution to the problem of longitude was in the end provided by John Harrison and his highly sophisticated marine chronometer, a version of which Halley had seen at a meeting of the Board of Longitude in 1737.

In 1720, the first year of his new position as Astronomer Royal, Halley published a paper on the observations of Jacques Cassini, the son of Jean Cassini, with whom Halley had stayed on his Grand Tour of Europe. He also published a paper on observing with a transit instrument. This was an observational instrument mounted in such a way that it could only point along the north-south line of the meridian and was used to observe and time the passage or transit of a star across the meridian. It was fitted with an eyepiece which had 'cross hairs', that often consisted of split strands of a spider's web.

Also in 1720, Halley made a contribution to cosmology, arguing against those who proposed that the universe was finite in extent. Halley's principal argument was that if the universe were finite it would imply that the universe had a central point. If this were so, then, according to the law of universal gravitation, all matter would be drawn towards this central point and the result would be the complete and utter destruction of the universe. Halley's arguments were not flawless, but they were an original and stimulating contribution to a field of study still in its early stages.

In 1729 Halley received the honour of being elected as a member of the Paris Academy of Sciences. It was also in this year that Queen Caroline, consort to King

George II, made an official visit to the Royal Observatory at Greenwich. We know that Halley had made a point of asking Sir Hans Sloane that he, Halley, might be personally introduced to Queen Caroline, possibly because he wanted to drop a hint to the Queen that the Astronomer Royal's salary was not as much as it might be. Queen Caroline must have taken the hint or at least been very impressed with Halley because after her visit she managed to persuade the King to grant Halley a pension. Technically, this was for the years he had spent as a naval officer from 1698–1701, but, as has been suggested, it might have been given to supplement Halley's small salary.

By at least the 1730s, the Fellows of the Royal Society were meeting for informal gatherings at a London coffee house before the official session in the evening. It is fairly probable and in keeping with his character that Halley was a key figure in these gatherings, from which evolved the Royal Society Club in 1743, one year after Halley's death. There was a report that, at the time when the meetings moved to Dean Court, Halley was obliged to eat fish because he had no teeth. He was, by this time, in his seventies.

In January 1736, Halley's wife Mary died. This must have been a great blow to Halley since their fifty-four-year-old marriage seems to have been relatively happy and harmonious, though it is fair to say that very little is known of their married life. Indeed, this personal tragedy may have precipitated the minor stroke that Halley suffered in the same year. Halley was, as a result, slightly paralysed in his right arm and had to have an assistant, Gael Morris, to help him in his work. It was also in 1736 that he drew up his will.

In 1741, Halley's health began to decline and matters could not have been improved by the news of the death of his son, Edmond, about whom nothing is known except that he was a naval surgeon.

Towards the very end of his life, Halley still continued his work. Then, according to the *Biographica Britannica*, 'his paralytic disorder gradually increasing, and thereby his strength wearing, though gently, yet continually, away he came at length to be wholly supported by such cordials as were ordered by his Physician', who was Richard Mead, physician to the King, 'till being tired of these he asked for a glass of wine, and having drank it expired as he sat in his chair without a groan on the 14th of January in the 86th year of his age'. He was buried at Lee, near Greenwich, next to his wife.

Edmond Halley is now most famous for his observations on the Comet that returns to the sun about every seventy-six years. This was not the case during his lifetime. He was a man of many talents – an astronomer, a scientist, a mathematician, and to that can be added his skills as a navigator, translator and even, from the verses he attached to Newton's great work, the *Principia*, a poet.

Halley was a geophysicist and did much valuable work in oceanography, geomagnetism and meteorology. His theories on trade winds and the earth's magnetism are not now accepted but they did provoke the necessary interest for more successful theories to be advanced. He improved scientific instruments and investigated methods of deep-sea diving. As a navigator, he devised a method of determining longitude at sea. He was an expert in the works of Ancient Greek geometry which he translated from texts written in Arabic. In short, Halley was interested in any problem or phenomenon that could be approached scientifically, ranging from social statistics, as in his comparison of the relative densities of population in London and Paris, to historical geography, as in his investigation into the time and place of Julius Caesar's landing in Britain. His calculations and tables on life expectancy instigated the beginnings of life insurance.

Let me add just one more little-known example of the diversity of his interests. In 1688, having been told by the supervisor of the Physic Garden, Chelsea, London (which is still going strong), that a plant covered by a glass cloche did well but if the cloche was covered by brown paper the plant died, Halley's interest was aroused. He set up a test whereby an opaque cloche covered a plant by day and a clear glass one by night. This would establish if it was light that was essential to the plant's life and not some other factor. In this he anticipated correctly by nearly a century the definitive work of two scientists, Ingenhousz and Priestley.

11

The Confusion over the Pronunciation of Halley's Name

The widespread contemporary pronunciation of Edmond Halley's name as 'Haley', rhyming with Bailey, is undoubtedly attributable in the main to a certain Bill Haley whose rock 'n' roll group, the Comets, had a substantial influence on popular music worldwide from the mid-1950s. Yet historically it can be shown that this pronunciation tallies with the entry in the register for Halley's marriage on April 20th, 1682 when his name was spelt 'Hailey'.

Then there are many who adopt what could be called a modern, logical way of pronunciation, making Halley rhyme with 'alley'. This is fortified by the recognised enunciation of 'hallelujah' or even the Hallé Orchestra and the pronunciation by many individuals bearing the name Halley.

Some doggerel written by H. H. Turner in connection with the return of the Comet in 1910 sets a suitable rhyming seal upon this version (even if he confused meteors and comets!).

> Of all the meteors in the sky,
> There's none like Comet Halley.
> We see it with the naked eye,
> and periodically.

This version has, however, been neatly offset by a verse from Dr John Mason which anticipates what I think is the correct pronunciation and the Comet's visibility in 1986.

> Of all the comets in the sky
> There's none like Comet Halley.
> We'll see it with the naked eye,
> But this time very *poorly*.

There is no doubt that variations in spelling at the time of Halley (1656–1742) leave little chance for the definitive pronunciation of his name to be pinpointed. But nevertheless I am a firm advocate in conjunction with Halley's biographer, Colin Ronan, and collector of his letters and archives, E. F. MacPike, in making Halley rhyme with 'Hawley'.

To take a logical peg on which to hang this assumption: everyone pronounces 'hall' as 'hawl'. Thus by adding the suffix '-ey' the name must become 'Hawley'. If one goes back to the variations in spelling his name when he was alive, one finds that he himself spelt it 'Halley'. Mis-spellings of his name in correspondence during his lifetime do include 'Hailey', 'Haley', 'Hally', 'Hayley', but most consistently 'Hawly' or 'Hawley'. Those best equipped to write his name in the manner in which in conversation they had heard him pronounce it, were those who had the best education and who were of sufficient standing to be authoritative. Here the evidence, I suggest, tips the scale in favour of Halley being pronounced as 'Hawley' – the way it was written in so many important documents of the time, documents which can still be seen in the British Public Record Office.

> The Queen [wrote the Earl of Nottingham to the Lord High Treasurer in October 1702] would have £200 advanced to Mr Edmund *Hawley* for his charges in going abroad on matters of importance.

And in another letter a week later, signed by Queen Anne, there is a safe conduct in Latin, which begins:

> *Edmd. Hawley arm. Lræ Salvi Conductus.*
> *Anna Dei Gratia Magnæ Britannicæ . . .*

However, in all honesty, I should report that later on the name is spelt 'Edmundo Halley'.

To end this controversy once and for all let me quote first Samuel Pepys, with whom Halley shared an interest in the Royal Navy. In one of Pepys's naval minutes it is recorded that:

> *Mr Hawley* was considered to have 'the most . . . competent degree . . . of the science and practice of navigation'.

In the contemporary authoritative tome, Luttrell's *Brief Relation of State Affairs*, the entry dated September 14, 1700 reads:

> Captain *Hawley*, the famous mathematician, is come to town from his expedition in the South Seas, and has given the Lords of the Admiralty an account of the Observations and discourses he has made there.

(The italics are mine.)

In setting up Halley's Comet Society, as there were no rules, no annual subscription and no committees, I felt we ought to have something by way of a common denominator, some unifying influence apart from Halley and his Comet which at the same time would provide some conversational one-upmanship. I therefore introduced the convention of our members pronouncing his name 'Hawley' as opposed to the other possibly more common versions. But you take your pick!

Perhaps a flippant footnote, from the *Griffith Observer* (October, 1984), best demonstrates the thorough confusion not only of pronunciation but also of the spelling of Halley's name, even today. It points out that a telescope merchant in Los Angeles changed his copy three times in newspaper advertisements: 'Get Ready for Haley's' became 'Hailey's' and then 'Hally's'. The *Griffith Observer* commented: 'with luck Comet Halley will have its name properly spelled by the time it actually starts to round the sun in late 1985.'

As a postscript let me say that we know from the archives and Halley's will that he spelt his name Edmond not Edmund – a style which probably derived from the way his name was spelt in Latin.

12

Believe it or not – a Pot-Pourri of Bizarre and Little-known Facts of Halley Lore

When the return of the Comet was first sighted on October 16th, 1982 through Mount Palomar's giant telescope, it was found to be beyond the orbit of Saturn and '*was as faint as the light from a single candle seen 27,000 miles away*'.

(NASA's *Comet Halley Handbook*, Second Edition)

In 1973 latterday soothsayers interpreted the heralded (and then disappointing) appearance of Comet Kohoutek as signalling the downfall of Richard Nixon's Watergate-battered administration.

(Dennis Overbye in *Discover* magazine, December 1981)

In their excitement to see if Halley was right or wrong, Voltaire is quoted as saying that no astronomer went to bed in 1758 which was the year when Halley had predicted the Comet's return. It gives hope to us all that despite such professional vigilance it was a rank amateur, a Saxon farmer called Palitzsch, who first "recovered" the Comet with an ordinary telescope about a month before the professionals spotted it.

(Sources numerous and *Reader's Digest* article by Blake Clark, December 1983)

It is said that 63 million years ago comet dust enveloped the earth and blocked light to such an extent that all vegetation died. This assumption is made to explain satisfactorily the sudden and otherwise inexplicable extinction of the Dinosaur family. Without foliage to eat they soon starved to death.

(Nigel Calder's *The Comet is Coming*)

A pair of luxury trains operated by the Central Railroad Company of New Jersey between 1929–43, and connecting New York and Atlantic City, were called 'The Blue Comet'. Each coach was named after a comet with Halley's taking pride of place.

(Library of Congress archives)

When alleged lack of money prevented NASA, the US space agency, from sending a space mission like those of Giotto (Europe), Russia and Japan, a certain Barbara Honegger of the White House staff, in conjunction with Stan Kent, President of Delta Vee Inc., a space activists organisation, had an ingenious idea to save the situation. They started up 'The Halley Fund' to attract donations of even a few dollars from US patriots (tax deductible, they pointed out). Then they would send up a special camera to photograph the Comet and have it not only paid for but made profitable to the investors from advertising and promotional gimmicks and from payments made from TV and cable stations who would be licensed to give live coverage of the fly-by. The costs and difficulties are enormous and at the time this book went to press it was not known to what extent their plans had matured.

(Advertisement in *Omni* magazine, July 1982. Geoffrey Golson's 'Comet Knowledge' article in TWA *Ambassador*, May 1982)

Membership of Halley's Comet Society with its distinctive '1986' logo on ties and medallions rocketed from one in 1975 to over 500 by 1983 – all by word of mouth. George Malcolm Thomson, a Fleet Street journalist of distinction, who actually saw Halley's Comet when but a lad in 1910, wrote pithily on April 22nd, 1983: 'The tie. Very many thanks. But I really deserve it, don't I! The only thing is, people see the figures on it and say "Oh, I thought you were older than that!" People are charming.'

(Halley's Comet Society archives)

The fact that Halley's Comet comes back at imprecise intervals ranging from just under seventy-five years to seventy-eight years (average about seventy-six years) has been the subject of considerable debate by astronomer/mathematicians over the centuries. One, Brady, in 1972 posed the theory that the pull of an as yet undiscovered planet ('Planet X') orbiting beyond Pluto, which is on the edge of our solar system, had to be envisaged. Later in 1972 Dr T. Kiang submitted a paper* refuting this and showing that it was the tripartite relationship between the sun, Jupiter and Halley's Comet which better accounted for the difference in time between each of its visits. What the layman will readily understand is that just as Newton recognised that what made his famous apple fall to earth was the pull of earth's gravity, so the Comet can be accelerated by the pull of a large planet according to its proximity. Just as important, however, is to note that if the distance between the Comet and the planet is sufficiently great, then lack of gravitational pull will slow the Comet down.

In January 1910, a new comet, called simply '1910 I', had the impertinence to turn up. It was of significance and magnitude to command wide public interest but it got its timing all wrong. Everybody was talking about and waiting for Halley's, so the newcomer was for all practical and popular purposes ignored.

('Yankie' article by Greg Stone, December 1973)

A comet of 1680 made a considerable impact which resulted in a dire warning to poultry breeders. A hen in Rome at that time, obviously emotionally disturbed at the sight of the comet or perhaps even inspired to commemorate it, produced an egg with a design of the comet on its shell. Several broadsheets were devoted to this feat and the egg was admired by the Pope and the Queen of Sweden no less.

*'The Cause of the Residuals in the Motion of Halley's Comet', paper for R. Astr. Soc. by T. Kiang.

The illustration painted in *Das Weltball* of the Comet markings on the egg laid by the inspired hen which coincided with the Comet of 1682.

Visual evidence was recorded in an old German woodcut of that period which shows the designs of the comet on the eggshell at that time. Well, would you believe it, in 1682 when Halley's Comet arrived there was this hen in Marburg (Germany) not to be outdone. It also laid an egg adorned with a comet design which was depicted in an old illustration reprinted in *Das Weltball* (April 1st, 1907). But now read on. The following is translated from the *Bulletin de la Société Astronomique de France*, August 1910:

'Madame Bouyard wrote on 17th May that one of her hens had laid an egg on the shell of which one could see very clearly a picture of a comet, a kind of big star with a tail.' Unfortunately no illustration of Madame Bouyard's hen nor its exciting egg seems to exist. In view of this tradition one can expect a plethora of comet eggs in 1986 now that the battery hen business is in full swing.

(Library of Congress, Washington DC, archives)

But the connection between hens, their eggs and the Comet must certainly be capped by this account in a Nevada (USA) newspaper published early May 1910.

COMET SHAPED EGG THE LATEST PRODUCTION

Halley's comet which is causing the superstitious so much worry and is also causing people to run around in their nighties and catch cold just to catch

> a glimpse of it is now interfering with the duties of the hens in Reno. If you don't believe it listen to this story concerning County Clerk Fogg and his pet hen. A few mornings ago County Clerk Fogg donning a bath robe went out into the back yard of his home about 3 o'clock to see Halley's comet. When he got out there he was puzzled to see his pet hen running around in the yard cackling and looking into the heavens. He watched the hen for a few minutes and then went into the house just as the hen retired to the hen house. Later that morning Mr Fogg went into the hen-roost and found in his pet hen's private nest an egg with a long tail on it. It was a comet egg and the hen it is believed was out in the yard to get a good look at the comet before laying this specimen for her master.

An interesting coincidence connecting Halley's of 1835 and 1910 was provided by Sir John Herschel's little refractor telescope. He used it to wave the 1835 comet goodbye when he observed it when it could last be seen in May 1836. It was used again as the first to welcome it some seventy-three years later. This is because it guided the larger telescope at the Helwan Observatory near the famous pyramids of Sakkara with which the comet was first photographed on August 24th 1909 on its last return.

(*Popular Astronomy*, October 1949)

Two Chinese researchers, Ren Zhenqui and Li Zhisen, predict a little ice age of cold winters and freezing disasters for the next two decades because on November 2nd, 1982 all the planets (except the earth) were grouped in a 'synod' on the opposite side of the sun. On the contrary says Arthur Mackins, an English weather-man from Bognor Regis, the Chinese have got it wrong. He says we should expect a major heatwave in the next years resulting from the return of Halley's Comet in 1986. Some of the hottest years on record have closely coincided with visits of Halley's Comet he says, just as the splendid summer of 1976 was due to West's Comet – or something.

(*The Times*, London, November 3rd, 1982)

Mark Twain was born on November 30th, 1835 only fourteen days after Halley's Comet of that year was closest to the sun. He died on April 21st, 1910 only one day after when the comet was closest to the sun. According to Albert B. Paine, Mark Twain's first biographer, as the author's heart attacks increased in 1909 he began to make preparations both private and public for his own demise. 'I came in with Halley's Comet in 1835,' he said to Paine. 'It is coming again next year and I expect to go out with it.' He did.

(Halley Comet Watch *Newsletter*, May–June 1982)

Another famous American author, James Thurber, was sixteen years old when Halley's was visible in 1910. In his book *My World and Welcome to it* he wrote humorously (being aware of the predictions of the disasters the Comet might bring), 'Nothing happened except that I was left with a curious twitching of my left ear after sundown and a tendency to break into a dog-trot at the striking of a match or the flashing of a lantern.'

(Halley Comet Watch *Newsletter*, May-June 1982)

In September 1984 the Soviet Union said that a Latvian folksong has helped astronomers trace sightings of Halley's Comet back to May 16th, 240 BC. It was stated that a certain Jan Kletniek, a surveyor and assistant professor at the Riga Institute, made the discovery.

(Source – Official News Agency TASS)

When the 'official Halley's Comet 2-Year Calendar' (covering 1985 and 1986) was published in the autumn of 1984 in the United Kingdom by W. H. Smith, the type chosen, quite unwittingly and coincidentally, for the text was called 'Horley'. At that time no reference in the calendar alluded to the confusion over the pronunciation of Halley, nor was the chapter in this book available ascribing authenticity to this version.

The 1986 space missions will hopefully sample the atmosphere of Halley's Comet but from an issue of the *Chicago Tribune* of 1910 it can be seen how this experiment was already anticipated with cheerful intemperance:

> At a special meeting of the Chicago General Committee for the Reception of Halley's Comet, Prof. Graham Taylor read a report from his Oxford colleague, Prof. Turner, stating that we shall be in the tail of the comet May 18, 1910 and if we wish to bottle some of the air that day we can hand a part of the comet down to our grandchildren. On behalf of the Committee, the Treasurer was instructed to purchase fifty dozen quarts of champagne for May 18. These, after being emptied, will be filled with Halley's best.

Published in 1679 a rare copy of a catalogue of southern stars observed by Halley when he was at St Helena fetched £6,500 at a Christie's Sale in October 1979.

(Report in London *Daily Telegraph*)

At the US Naval Observatory, Washington DC, a telescope is being dismantled to be sent all the way to a northern location on the South Island of New Zealand. It will be re-assembled to photograph the tails of the departing Comet after perihelion in the spring of 1986. But the small dark-room alongside the telescope in Washington in which photographic plates are changed will not go with it. It was originally Franklin Roosevelt's small private elevator.

(Halley's Comet Society archives)

Two cartoons showing early, if not premature, anticipation of Halley's Comet come from ***Washington Post*****, March 6th, 1981 and the London** ***Evening Standard*** **of June 26th, 1978 respectively.**

A cable TV programme in the USA featuring Orson Welles on the prophecies of Nostradamus, mentioned the disasters which would accompany the return of the 1986 Comet. The verses Nostradamus wrote which are said to presage disasters and profound changes in world history are Quatrain 43:

> During the time when the Hairy
> Star is apparent,
> The Three Great Princes shall be
> made Enemies,
> Struck from heaven, place quaking
> Earth,
> Arne, Tiber, full of surges,
> Serpents cast upon the shore.

Then Quatrain 62:

> Mabus shall come, and soon after shall die,
> Of people and beasts shall be a horrible
> destruction,
> Then on a sudden the vengeance shall
> be seen,
> Blood, sand, thirst, famine, when
> the Comet shall run.

There is always great argument about the accuracy and ambiguities of the many verses he wrote, but there is no doubt that the dire warnings he conveys for the Comet are right in line with the alarm and despondency with which people viewed the Comet throughout history.

In 1910 the zookeepers at Chicago's Lincoln Park zoo recalled that there had been an epidemic of kangaroo deaths in Australia when Halley's Comet appeared previously, in 1835. To avoid any risk of a repetition they moved all the young kangeroos indoors.

(Jay Maeder, *Miami Herald* writer)

An astronomy laboratory called Astro is being prepared by NASA to be sent up in the cargo bay of the now well-tried American Shuttle. Three scientists have been selected for these Astro missions which will use three telescopes and two cameras for a long-range, seven-day vigil of the comet in early March 1986 at the same time as the five unmanned interception craft from the Soviet Union, Japan, and Europe (Giotto) are closing in on the Comet.

(*Spaceflight*, volume 26, November 1984)

Professor Chandra Wickramasinghe, Head of the University College of Wales Astronomy Department, warns that the comet could bring with it influenza viruses preserved in the deep-freeze of outer space. He explained that the time lag between the Comet's last appearance and the terrible 'flu epidemics afflicting the world in 1910 and later from 1916 to 1918 was caused by the length of time it takes for microbes to drift down from the outer atmosphere.

(*Western Mail*, Cardiff, January 1985)

SIGHT THAT SHOCKED FLYING MONK

Mr Max Woosnam, an engineer of Bristol House, Malmesbury, reveals as a result of his interest in the history of the local abbey, that Elmer, a Saxon monk, saw the Comet twice during his long life from 980–1080 AD. The historian monk, William of Malmesbury, wrote of Elmer (which he spelt 'Eilmer') that when he saw the Comet 'He sagely cried – thou art come . . . I have seen thee before, but now I hold thee much more terrible, threatening to hurl destruction on this land.'

Elmer was famous for his feat of fastening wings to his hands and feet and launching himself from the top of a high tower. Assisted by a powerful breeze he flew over two hundred yards before crashing. He broke his legs and was lame thereafter.

(*Wiltshire Gazette*, January 3rd, 1985)

When the comet was last here in 1910 income tax was six old pence in the pound, but only for incomes over £3,000 (£120,000 by 1985 standards). A made-to-measure suit cost £1.50, and a week in Paris with full board only £4. A solidly-built detached house with several bedrooms in the London suburbs would fetch £500.

Daily Express (January 28th, 1985)

Under the heading of HALLEY'S COMET, REASON FOR THE SAGA OF THE AIR BATTLE OF THE SPIRITS AFTER THE HUNNEN SLAUGHTER Professor Hennig of Dusseldorf, seeing the Comet in 1910, wrote in a German periodical about the persistence of the legend that after this slaughter the 'rage of the fighters was so great that even the spirits of the fallen carried on fighting in the air'.

The saga of the spirit battle, said Professor Hennig, had its origin in a record dated 470 which, translated from the Greek, went as follows:

'But the most surprising event, as it is related, was the following:
As the fighters fell, after giving up their bodies, the warriors rose anew in their souls. For a further three whole days and nights they fought, entangled with the living with enraged hands. The contours of the fighting spirits with sounding of weapons, were seen and heard. And other such ancient apparitions of battles are still to be seen until today.'

Professor Hennig then worked out that the Comet's perihelion in 451 which produced an apparition of 'tremendous size' coincided with the date of this 'world-historic' battle and concluded that the legend of the spirits of the fallen continuing to fight in the heavens could only be thus explained.

Let me end this chapter with a cheerful (and possibly apocryphal) Comet story:

OPERATION HALLEY'S COMET

A colonel issued the following directives to his executive officers:

'Tomorrow evening at approximately 2000 hours Halley's Comet will be visible in this area, an event which occurs only once every seventy-five years. Have the men fall out in the battalion area in fatigues, and I will explain this rare phenomenon to them. In case of rain, we will not be able to see anything, so assemble the men in the theatre and I will show them films of it.'

Executive Officer to Company Commander:

'By order of the Colonel, tomorrow at 2000 hours, Halley's Comet will appear above the battalion area. If it rains, fall the men out in fatigues, then march to the theatre where this rare phenomenon will take place, something which occurs only once every seventy-five years.'

Company Commander to Lieutenant:

'By order of the Colonel in fatigues at 2000 hours tomorrow evening, the phenomenal Halley's Comet will appear in the theatre. In case of rain, in the battalion area, the Colonel will give another order, something which occurs once every seventy-five years.

Lieutenant to Sergeant:

'Tomorrow at 2000 hours, the Colonel will appear in the theatre with Halley's Comet, something which happens every seventy-five years. If it rains, the Colonel will order the comet into the battalion area.'

Sergeant to Squad:

'When it rains tomorrow at 2000 hours, the phenomenal seventy-five-year old General Halley, accompanied by the Colonel, will drive his comet through the battalion area theatre in fatigues.'

13

Comets in Literature and History

The comet has long appealed to writers, historians and thinkers throughout history. Shakespeare, Pepys, Luther, Milton and Tolstoy have each had something to say about comets in their works. The following section has been arranged in chronological order so that it is possible to trace a shift in popular attitude towards comets as science has worked to dispel the cloud of superstitions that has surrounded them.

From ancient times up to the Renaissance comets were almost always associated with doom and disaster and especially with the death of a king or ruler. Although there is no evidence to support this connection, the sentiment expressed by Calpurnia in Shakespeare's lines from *Julius Caesar*, 'When beggars die there are no comets seen,/The Heavens themselves blaze forth the death of princes,' occurs again and again in this or a similar form up to the time of Halley and even after.

Interestingly, though, Shakespeare also uses the image of the comet in *Henry VI Part I* to invoke a *positive* change of fortune and so extends the comet's usual associations of disaster to those of a favourable change in circumstances.

By the beginning of the eighteenth century, Newton and Halley had done enough work on comets to show that they were not so much freakish, fiery ghosts arising out of the blue, but celestial bodies orbiting the sun. This scientific progress eventually percolated down to the man in the street. The result was that the comet became to some extent freed from its associations with disaster and writers could look to other aspects for images and symbols. Tolstoy, for example, in a sublime passage from *War and Peace*, emphasises the comet's awe and mystery, while Thomas Hardy points out its once-in-a-lifetime aspect.

It is perhaps not inappropriate here to begin with a poem written especially for the 1985–6 return of Halley's Comet. In only a dozen lines it encapsulates the majesty and mystery of this great enigma of our heavens. It was written by Mary Wilson in December 1984 at my request.

TO HALLEY'S COMET

Your sparkling hair streams back across the skies,
And moonbeams shrivel in your fiery breath;
Your glory is reflected in our eyes –
Yet do you bring us pestilence and death,
Parched burning summers, freezing winter cold?
What is your gift to us, bright orb of gold?

The space-ship, bearing greetings, as a friend
Is flying out to guide you on your way;
And from your lonely circuit without end
Perhaps your radiance brings a better day,
And, like the star which heralded a Birth –
A hope for peace and happiness on earth.

Mary Wilson

The Halley's Comet Society, which is proud to have Mary Wilson and Lord Wilson of Rievaulx as founder members, unites in thanking her.

The Bible: There is a possible reference to a comet in I Chronicles 21:16. This is strengthened by the fact that, in a similar fashion, the Jewish historian Josephus referred to Halley's Comet in AD 66 as hanging over Jerusalem like a sword. It is the comet's tail, sweeping behind the nucleus, that can make a comet seem like a giant scimitar in the sky. The Bible reference is as follows:

> And David lifted up his eyes and saw the angel of the Lord stand between the earth and the heaven, having a drawn sword in his hand stretched out over Jerusalem . . .

Virgil (70–19 BC): Roman poet who wrote the *Aeneid*, the greatest epic poem written in Latin, and the *Georgics*, a poem about farming. The first extract, from the *Aeneid*, describes the helmet of Aeneas, the hero of the book, in a manner reminiscent of the way Homer describes the helmet of Achilles in the *Iliad*:

The crest on Aeneas's helmet blazed,
Flames poured from the plume on top,
And the gold centre of his shield
Threw out vast fiery flashes
As when, on a clear night,
A comet glows a sinister blood-red.

Comets themselves have not been known to be a blood-red colour though it is possible that a comet's light might be altered by the earth's atmosphere.

The second extract, from the *Georgics*, describes the atmosphere at the time of the death of Julius Caesar. It was believed by some that the comet of 44 BC was in fact the departed soul of Caesar, rising to the heavens:

. . . At this time
Threatening signs never stopped appearing
In entrails, that gave no reassurance;
Nor did blood stop flowing from the wells.
Throughout the night, high cities
Echoed with the howling of wolves.

At no other time did more lightning-
Flashes streak the clear skies,
Nor so often did ominous comets blaze.

Seneca (*c*.3 BC–AD 65): Roman philosopher and writer whose death occurred a year before the return of Halley's Comet in AD 66. Seneca was accused by Nero, to whom he was once tutor, of taking part in the Piso conspiracy to assassinate Nero and was forced to commit suicide. The two extracts come from *Naturales Questiones*. The first one indicates that Seneca thought that comets were actual celestial bodies that moved in space and not phenomena of the earth's upper atmosphere – as Aristotle (384–322 BC) took them to be. The 'deified Julius' refers to Julius Caesar:

> There is nothing to suppose that the comet seen during Claudius's reign was the same one that we saw when Augustus was in power; nor that the comet which appeared during Nero's reign and diminished the comet's reputation for bringing bad fortune, resembled the one that blazed into view at dusk on the day of the games for Venus Genetrix, after the deified Julius died.

It is interesting to note in the extract above that Seneca even considered the possibility of a comet returning, some 600 years before Halley was able to predict successfully that the 1682 comet would return in 1758.

In the second extract the comet referred to could in fact be a nova, a star that suddenly increases its brightness dramatically, because of internal explosions, and then diminishes in brightness over a longer period:

> After the Syrian king, Demetrius, died [in 151 BC] . . . shortly before the Achaean war, a comet blazed forth; to start with it was a ruddy, burning disc so brilliant that it turned night into day. Then gradually its size and brilliant light died away.

Pliny the Elder (AD 23–79): Roman soldier and writer. Pliny was the author of *Naturalis Historia*, a book on which Halley wrote a paper in 1691. Pliny here refers to the etymology of the word 'comet', which comes from the Greek meaning 'long-haired', in accordance with the appearance of a comet's long, sweeping tail.

> And there are stars that are suddenly born in the sky itself. There are several kinds of these stars. They are called by the Greeks 'comets', while we call them 'long-haired' stars . . .

Silius Italicus (AD 26–101): Roman poet who wrote *Punica*, the longest poem in Latin with 12,200 verses. It relates the events of the Second Punic War between the Romans and the Carthaginians under Hannibal. This extract from the *Punica* describes the eerie atmosphere at Cannae before the battle in which the Romans were decisively defeated by Hannibal in 216 BC. Silius makes it clear what the appearance of a comet stood for at that time:

> Again and again a screech-owl haunted the gates of the camp. Dense clouds of bees enveloped the quivering standards; more than one comet, the destroyer of kings, with its shimmering tail glowed a deadly red colour.

Josephus (AD 37–?): Jewish priest and Pharisee who later became a Roman citizen. He wrote a history of the Jewish war between the Jews and the Romans, which broke out in AD 66, the year Halley's Comet returned. Josephus described Halley's Comet as being stretched out 'like a sword' over the city of Jerusalem. This tallies with the possible reference to a comet in the Bible (I Chronicles 21:16). Josephus saw the comet as portending disaster for the Jews: in AD 70, Vespasian and Titus sacked Jerusalem and destroyed the Temple. It was, by the way, Vespasian who, after seeing the comet of AD 79, is said to have commented: 'This hairy star does not concern me; it menaces rather the King of the Parthians, for he is hairy, while I am bald.' Actually Vespasian did die not long afterwards.

Lucan (AD 39–65): Roman poet who wrote *Pharsalia*, describing the struggle between Julius Caesar and Pompey. Lucan, like Seneca, was forced to commit suicide in AD 65, a year before the return of Halley's Comet, after being implicated in the Piso conspiracy to assassinate Nero. Here he describes the portents that occurred before Caesar reached Rome:

> The dark nights were witnesses to unknown stars,
> The sky burnt with fires,
> Shooting stars zipped across the vast heavens
> And the long hair of the frightening star – the comet
> That brings change to kings.

Tacitus (AD 56–*c.*117): Roman historian. His work, the *Annals*, covers the period of history between the death of Augustus in AD 14 and the death of Nero in AD 68. Tacitus describes the brutal way in which Nero dealt with the ominous appearances of comets:

> At the end of the year, news spread of portents announcing that disasters were about to happen – an unprecedented number of lightning flashes and a comet (a portent which Nero always averted with the blood of noblemen).

Here, Tacitus describes in a very matter of fact way the powerful effect of a comet on the Roman people:

> Meanwhile, a comet, popularly thought to be a phenomenon portending change in governments, blazed brightly. And so, as if Nero had already been deposed, there was general speculation as to who would take power next.

Suetonius (*c.* AD 69–*c.* 160): Roman historian who wrote *Lives of the Caesars*. He would have seen the return of Halley's Comet in AD 141. In his life of Nero, Suetonius describes, with more detail than Tacitus, how Nero dealt with the threat he felt came from a comet:

> A comet, which is a phenomenon popularly supposed to herald the death of the highest rulers, began to rise in the sky on successive nights. Nero was troubled by this and, hearing from the astrologer Balbillus that kings were accustomed to avert such portents by murdering someone of noble blood, thus deflecting these portents from themselves on to the heads of noblemen, he [Nero] determined to kill all the most distinguished nobles . . .

Bede (AD 673–735): English historian who lived at the monastery at Jarrow for most of his life. Bede is here describing the appearance of two comets in AD 729:

> In the year of our Lord 729, two comets appeared around the sun, striking terror into all who saw them. One comet rose early and preceded the sun, while the other followed the setting sun at evening, seeming to portend awful calamity to east and west alike. One comet was the precursor of day and the other of night, to indicate that mankind was menaced by evil at both times. They appeared in the month of January, and remained visible for about a fortnight, bearing their torches northward as though to set the welkin aflame.

Anglo-Saxon Chronicle. This chronicle was compiled by monks in England up to the middle of the twelfth century. The extract here refers to the celebrated return of Halley's Comet in 1066, the year of the Battle of Hastings when King Harold was defeated by William of Normandy. It is said that while Harold took the Comet as presaging disaster, William saw it as an omen of good luck. Halley's Comet was subsequently commemorated on the Bayeux Tapestry where it looks rather like a shuttlecock:

> In this year King Harald came from York to Westminster at Easter, which was after the mid winter in which the King died. Then was seen over all England such a sign in the heavens as no man ever before saw; some say it was the star Cometa, which some men call the haired star, and it first appeared on the eve of Litania-Major, the 8th of the Kalends of May, and so shone all the seven nights.

Dante Alighieri (1265–1321): Italian poet born in Florence. Dante would have seen the return of Halley's Comet in 1301 as would his fellow Florentine, Giotto, the painter who used a representation of the Comet to depict the Star of Bethlehem in his Nativity scene. In the *Paradiso*, the last third of Dante's great work *The Divine Comedy*, here translated by Longfellow, Beatrice, the poet's guide, has just spoken:

Thus Beatrice: and those souls beatified
Transformed themselves to spheres on steadfast poles,
Flaming intensely in the guise of Comets.

Martin Luther (1483–1546): Leader of the Reformation in Germany. Luther would have been forty-eight years old when Halley's Comet returned in 1531. It is interesting to note the sort of human qualities that Luther here ascribes to a comet:

> A comet is a star that runs not being fixed like a planet, but a bastard among planets. It is a haughty and proud star engrossing the whole element, and carrying itself as if it were there alone.

Tasso Torquato (1544–95): Italian poet born in Sorrento. Tasso was author of the epic poem *Jerusalem Delivered*, here translated by Wiffen, which describes the events of a Christian crusade. He faithfully follows the traditional fear-and-trembling associations of the Comet:

As with its bloody locks let loose in the air,
Horribly bright, the Comet shows whose shine
Plagues the parched World, whose looks the Nations scare,
Before whose face States change, and Powers decline,
To purple Tyrants all, an inauspicious sign.

Edmund Spenser (1552–99): English poet. Spenser wrote the *The Faerie Queene*, from which this extract comes. He uses the image of a comet's tail to describe the hair of a girl in flight. He does not give the gloomy associations of a comet directly, but in a more guarded fashion, via 'the sage wisard . . . as he has red':

Still as she fled, her eye she backward threw,
As fearing evill, that persewd her fast;
And her fair yellow locks behind her flew
Loosely disperst with puffe of every blast:
All as a blazing starre doth farre outcast
His hearie beames, and flaming lockes dispred,
At sight whereof the people stand aghast:
But the sage wisard telles, as he has red,
That it importunes death and dolefull drerihed.

William Shakespeare (1564–1616): Shakespeare would have seen Halley's Comet in 1607. As well as probably the most well-known quotation on comets, spoken by Calpurnia in *Julius Caesar*, and referred to above, in *Henry VI Part I* Shakespeare takes a different view following the death of King Henry V:

Hung be the heavens with black, yield day to night;
Comets importing change of Times and States,
Brandish your crystal tresses in the sky,

And with them scourge the bad revolting stars,
That have consented unto Henry's death . . .

John Donne (1572–1631): English poet who became Dean of St Paul's. In this extract from his poem *The First Anniversary*, Donne is contrasting the brevity of human life with what it used to be in the days of Methuselah, who lived to be 969 years of age. Donne calls a comet a 'slow-paced star' and interestingly makes reference to the comet returning to be seen again – a phenomenon that would be successfully predicted by Halley, eighty-five years after Donne's poem was published in 1611.

There is not now that mankind, which was then,
When as the sun, and man, did seem to strive,
(Joint tenants of the world) who should survive.
When stag, and raven, and the long-lived tree,
Compared with man, died in minority;
When, if a slow-paced star had stol'n away
From the observer's marking, he might stay
Two or three hundred years to see 't again,
And then make up his observation plain.

John Milton (1608–74): English poet. Milton was a pupil at St Paul's School where Halley later went. In this first extract from his great poem *Paradise Lost*, Milton captures the cosmic rage of Satan with the comet simile. 'Ophiucus' is the name of a constellation:

Incens'd with indignation Satan stood
Unterrified, and like a comet burn'd
That fires the length of Ophiucus huge
In th'arctic sky, and from his horrid hair
Shakes pestilence and war . . .

In this second extract, Milton uses the image of a comet specifically to convey the heat of the blazing sword. But he must also have had in mind the visual image of a comet resembling a drawn sword – the image used by Josephus (see page 142) and possibly in the Bible, I Chronicles 21:16 (see page 140).

The brandished sword of God before them blazed
Fierce as a comet; which with torrid heat . . .
Began to parch that temperate clime . . .

Andrew Marvell (1621–78): English poet. Marvell, in his poem 'The Mower to the Glow-worms', alludes to the comet's gloomy associations and wittily describes the glow-worms as the equivalent harbingers of doom in their world:

Ye country comets, that portend
No war nor prince's funeral
Shining unto no other end
Then to presage the grasses fall.

Samuel Pepys (1633–1703): Famous for his diary, which he began in 1660. The first three extracts from his diary refer to the comet of 1664:

> December 17th, 1664: Mighty talk there is of this Comet that is seen a'nights; and the King and Queene did sit up last night to see it, and did, it seems, and to-night I thought to have done so too; but it is cloudy, and so no stars appear. But I will endeavour it.
>
> December 21st, 1664: My Lord Sandwich this day writes me word that he hath seen (at Portsmouth) the Comet, and says it is the most extraordinary thing he ever saw.
>
> December 24th, 1664: I saw the Comet, which is now, whether worn away or no I know not, but appears not with a tail, but is larger and duller than any other star, and is come to rise betimes, and to make a great arch, and is gone quite to a new place in the heavens than it was before: but I hope in a clearer night something more will be seen.

The fourth extract from Pepys's diary indicates that Robert Hooke, a Fellow of the Royal Society, must have been fully aware by 1665 of the possibility of a comet returning to the sun. This is interesting because when Halley wrote to Hooke from Paris in 1680 and mentioned Cassini's theory that the recent comet might have been the same one that appeared in 1665 and 1577, Halley adds, 'I know you will with difficulty Embrace this Notion of his . . .' This suggests that Halley did not know of Hooke's earlier thoughts on the matter or else that Hooke had radically altered his views between 1665 and 1680:

> March 1st, 1665: To Gresham College, where Mr Hooke read a second very curious lecture about the late Comet; among other things proving very probably that this is the very same Comet that appeared in 1618, and that it will appear again, which is a very new opinion; but all will be in print.

Daniel Defoe (1660?–1731): English writer and author of *Robinson Crusoe.* Defoe wrote a historical fiction describing life in London during the Great Plague called *A Journal of the Plague Year.* Here he describes the comet of 1664 – the comet which Samuel Pepys mentions in his diary above:

> It passed directly over London so that it was plain that it imported something peculiar to the city alone . . . it was of a faint, dull languid colour, that its motion was very heavy, solemn and slow, and it accordingly foretold a heavy judgement, slow but severe, terrible and frightful, as was the Plague.

Jonathan Swift (1667–1745): Satirical writer born in Dublin, author of *Gulliver's Travels*. Swift's terse comment on comets shows his lack of reverence for the superstitious connotations associated with them:

> Old men and comets have been revered for the same reason; their long beards and their pretences to foretell events.

William Congreve (1670–1729): Irish dramatist who was a fellow student of Swift. In his play *The Mourning Bride* (in which occur the famous lines 'Heaven has no rage, like love to hatred turned,/Nor hell a fury, like a woman scorned') Congreve uses the comet as a metaphor for a state of disharmony. The traditional comet associations of doom and disaster have here been replaced by a certain feeling of futility:

> Nature to each allots his proper sphere,
> But that forsaken, we like comets err.
> Toss'd thro' the void, by some rude shock we're broke,
> And all our boasted fire is lost in smoke.

Joseph Addison (1672–1719): English essayist and founder, with Richard Steele, of the *Spectator*. The 'great Ferment' that Addison refers to was the War of the Spanish Succession which had broken out at the turn of the century:

> According to Sir Isaac Newton's calculations, the last Comet that made its appearance in 1680, imbibed so much Heat by its Approaches to the Sun, that it would have been two thousand times hotter than red hot Iron, had it been a Globe of that metal; and that supposing it as big as the Earth, and at the same distance from the Sun, it would be fifty thousand years in cooling, before it recover'd its natural Temper. In the like manner, if an *English* Man considers the great Ferment into which our Political World is thrown at present, and how intensely it is heated in all its Parts, he cannot suppose that it will cool again in less than three hundred years!
>
> (*Spectator*, June 26th, 1711)

Alexander Pope (1678–1744): English poet. In his poem *Essay on Man*, Pope speculates on the limits of man's knowledge in light of the scientific revolution brought about by Sir Isaac Newton. The 'he' refers, in fact, to Newton:

> Could he, whose rules the rapid comet bind,
> Describe or fix one movement of his mind?
> Who saw its fires here rise, and there descend,
> Explain his own beginnings, or his end?

Edward Young (1683–1765): English poet who was born a year after Halley's Comet returned in 1682 and so just missed seeing it twice. Young is the author of *Night Thoughts*, a poem of some 10,000 lines, whose first publication was in 1742, the year Halley died. Young, here, combines traditional imagery of the comet as a fear-inducing phenomenon along with, in the last two lines, contemporary scientific knowledge:

> Hast thou ne'er seen the comet's flaming light?
> Th'illustrious stranger passing, terror sheds
> On gazing nations, from his fiery train
> Of length enormous, takes his ample round
> Through depths of ether; coasts unnumber'd worlds,
> Of more than solar glory; doubles wide
> Heaven's mighty cape; and then revisits earth,
> From the long travel of a thousand years.

James Thomson (1700–48): Scottish poet, author of *The Seasons*. This extract from 'Summer' is similar in tone and imagery to what Edward Young wrote above in *Night Thoughts*. It seems that Thomson knew that the closer a comet gets to the sun, the faster it moves:

> Lo! from the dread immensity of space
> Returning, with accelerated course,
> The rushing comet to the sun descends:
> And as he sinks below the shading earth,
> With awful train projected, o'er the heavens,
> The guilty nations tremble.

Christopher Smart (1722–71): English poet. Smart would have been thirty-six years old when Halley's Comet returned in 1758. The extract is from 'Song to David', a poem that praises the author of the Psalms. The comet is here something to be praised for its beauty and not feared as a harbinger of disaster:

> Glorious the sun in mid-career;
> Glorious th'assembled fires appear;
> Glorious the comet's train.
> Glorious the trumpet and alarm;
> Glorious th'almighty stretched-out arm;
> Glorious th'enraptured main.

William Wordsworth (1770–1850): English poet. Wordsworth would have been sixty-five years old when Halley's Comet returned in 1835. The two extracts are from the original quarto of his *Descriptive Sketches* and the revised version of the same lines. With the omission of the comet image in the revised version, the dramatic impact of the verse has been significantly reduced:

> *Original quarto, 1793*
> The measured echo of the distant flail,
> Winded in sweeter cadence down the vale;

A more majestic tide the water roll'd
And glowed the sun-gilt groves in richer gold:
–Tho' Liberty shall soon; indignant, raise
Red on his hills his beacon's comet blaze . . .

Revised version
The measured echo of the distant flail
Wound in more welcome cadence down the vale;
With more majestic course the water rolled,
And ripening foliage shone with richer gold.
–But foes are gathering–Liberty must raise
Red on the hills her beacon's far-seen blaze . . .

Robert Southey (1774–1843): English poet who was Poet Laureate from 1813 to 1843. He was a friend of Wordsworth. These fast-moving lines are from 'St Antidius, the Pope and the Devil':

He ran against a shooting star,
So fast for fear did he sail,
And he singed the beard of the Bishop
Against a comet's tail;

Lord Byron (1788–1824): English poet and leading figure of the Romantic movement. In these lines from 'Churchill's Grave', Byron contrasts the energy of Churchill when he was alive with the pitiful state of his grave:

I stood beside the grave of him who blazed
The comet of a season, and I saw
The humblest of all sepulchres, and gazed
With not the less of sorrow and of awe
On that neglected turf and quiet stone . . .

John Keats (1795–1821): English poet who wrote the famous 'Ode to a Nightingale'. The extract is from a letter from Keats to Fanny Brawne, the woman he was in love with, dated July 8th, 1819:

> I have seen your Comet, and only wish it were a sign that poor Rice would get well whose illness makes him rather a melancholy companion: and the more so as to conquer his feelings and hide them from me, with a forc'd Pun.

Alfred, Lord Tennyson (1809–92): English poet. Tennyson would have been twenty-six years old when Halley's Comet returned in 1835. These lines from *The Lady of Shalott* are reminiscent of Virgil's description of the helmet of Aeneas. It is interesting to note that Tennyson has here confused a 'meteor' with a 'comet'. A meteor is, of course, a shooting star – a sudden streak of light in the night sky. Here, the fact that the 'meteor' is 'bearded' and the verbs describing it – 'trailing' and 'moves' – give the sense of a slowish movement, clearly indicates that Tennyson had a comet in mind:

All in the blue unclouded weather
Thick-jewell'd shone the saddle-leather,
The helmet and the helmet-feather
Burn'd like one burning flame together,
As he rode down to Camelot.
As often thro' the purple night,
Below the starry clusters bright,
Some bearded meteor, trailing light,
Moves over still Shalott.

The next two extracts come from Tennyson's verse play *Harold*, about King Harold who was defeated at the Battle of Hastings. The play opens with the English nobles and peasantry terrified by the sight of a great comet (it was, of course, Halley's Comet in 1066) in the skies:

It glares in heaven, it flares upon the Thames,
The people are as thick as bees below,
They hum like bees, – they cannot speak – for awe . . .

Later, Archbishop Stigand is asked what he thinks the comet signifies:

Not I. I cannot read the face of heaven;
Perhaps our vines will grow the better for it.

It is interesting to note that the notion of a comet affecting vines was well-known in the early nineteenth century after the year 1811 when the great Comet of that year was held responsible for the excellent wine harvest in Portugal. For years afterwards reference was made to the 'Comet wine' of 1811. The notion was based on the unsupported theory that the comet caused the right sort of hot weather for the vines to prosper.

Lastly, Archbishop Stigand advances to Harold and points up towards the comet. In Harold's answer to Stigand, Tennyson again wrongly calls 'comets' 'meteors':

Stigand: War there, my son? Is that the doom of England?
Harold: Why not the doom of all the world as well?
For all the world sees it as well as England.
These meteors came and went before our day,
Not harming any: it threatens us no more
Than French or Norman. War? The worst that follows
Things that seem jerk'd out of the common rut
Of Nature is the hot religious fool,
Who, seeing war in heaven, for heaven's credit
Makes it on earth . . .

Count Tolstoy (1828–1910): Russian novelist. Tolstoy died in the November of the year Halley's Comet returned, and so would have seen it twice. He was the author of *War and Peace*, the great epic novel from which this extract comes. Here, the comet signifies an inner feeling of calm and harmony experienced by Pierre:

> It was clear and frosty. Above the dirty ill-lit streets, above the black roofs, stretched the dark starry sky. Only as he gazed up at the heavens did Pierre cease to feel the humiliating pettiness of all earthly things compared with the heights to which his soul had just been raised. As he drove out on to Arbatsky Square his eyes were met by a vast expanse of starry black sky. Almost in the centre of this sky, above the Prichistensky Boulevard, surrounded and convoyed on every side by stars but distinguished from them all by its nearness to the earth, its white light and its long uplifted tail, shone the huge, brilliant comet of the year 1812 – the comet which was said to portend all manner of horrors and the end of the world. But that bright comet with its long luminous tail aroused no feeling of fear in Pierre's heart. On the contrary, with rapture and his eyes wet with tears, he contemplated the radiant star which, after travelling in its orbit and inconceivable velocity through infinite space, seemed suddenly – like an arrow piercing the earth – to remain fast in one chosen spot in the black firmament, vigorously tossing up its tail, shining and playing with its white light amid the countless other scintillating stars. It seemed to Pierre that this comet spoke in full harmony with all that filled his own softened and uplifted soul, now blossoming into a new life.

Thomas Hardy (1840–1928): English novelist and poet. Hardy was author of *Far From the Madding Crowd* and *Tess of the d'Urbervilles*. In his novel *Two on a Tower*, Hardy uses the excitement of a comet appearing to stimulate the recovery of the hero, an astronomer, who is seriously ill. In the poem quoted here, Hardy stresses the once-in-a-lifetime aspect of a comet:

The Comet at Yell'ham

It bends far over Yell'ham Plain,
  And we, from Yell'ham Height,
Stand and regard its fiery train,
  So soon to swim from sight.

It will return long years hence, when
  As now its strange swift shine
Will fall on Yell'ham; but not then
  On that sweet form of mine.

Gerard Manley Hopkins (1844–89): English poet. The extract is from his journal dated July 1874. The language is characteristically poetical and his likening the comet to a shuttlecock reminds one of the depiction of Halley's Comet on the Bayeux Tapestry, where it looks very much like a shuttlecock:

> The comet – I have seen it at bedtime in the west, with head to the ground, white, a soft well-shaped tail, not big: I felt a certain awe and instress, a feeling of strangeness, flight (it hangs like a shuttlecock at the height, before it falls), and of threatening.

14

Halley Versifying – 1910 Style

Comets have inspired poets and versifiers since man first recorded his religious, romantic and scientific thoughts (as we have seen in the previous chapter). But by the time of the 1910 appearance, much more was known about the movements and make-up of heavenly bodies. Scientific knowledge and deduction, combined with the strong humanist influence of the period and the familiarity born of a score of visits, robbed the Comet of much of its reverence and mystique. In the circumstances it was all of a piece that the exuberant Americans should provide most of the versifying, given that in many parts of the United States excellent visibility could be enjoyed – a facility less often shared by the northern Europeans and the British.

Even so a nervous gaiety could be detected here and there, the kind of spirit in which in the Second World War people joked about imminent bombing, an attempt to laugh about – even trivialise – what remained a largely unknown quantity that could still possibly carry some unpleasant consequences, however harmless previous appearances might have been.

Certain it is that the Comet rhymesters of 1910 did not soar into the empyrean where verse becomes transformed into poetry. At best the effusions were reasonably imaginative, reasonably well-constructed rhymings, but for the most part they were doggerel, often by intention.

Here is a selection:

Oh, you comet!
De troubles we has had
Wif chicken coops an' melon vines
Is sumpin' mighty sad.
We don't know what you's after,
We acks cheerful as we can,
But we mus' hab our suspicions
Of de midnight lantern man.

Philander Johnson, Washington *Star*

Of gases all compounded,
Forever chasing far,
And yet engaged in doing
Naught in particu-lar;
The solar system putting

At sixes and at sevens –
Hail and farewell, O comet,
Thou Roosevelt of the heavens!

The Public

No more politicians,
No more tariff schemes,
No more trust conditions,
No more quick-rich dreams!
Bang! annihilation!
Smash! we fly to bits!
There's some consolation
If the comet hits!

Paul West, New York *World*

That things have been at sixes and at sevens,
That something has gone wrong with Nature's laws,
Is due to you, newcomer to the heavens –
You are the cause!

Is England nice and prosperous? Far from it!
Look at the deadlock in her politics!
And you have brought about, unwelcome comet,
Our present fix.

And if today this modest poet's singing
Lacks sparkle, effervescence, grit and 'go',
It's all because to our old earth you're bringing
Your tail of woe!

FJC, London *Sunday Times* (following widespread reports that disasters were being ascribed to the influence of Halley's Comet)

Oh, Newton! Sky explorers all we are,
Halley, for us the light you traced is dawning.
Lo! the whole world hangs breathless on a star!
(Excuse my yawning).

Oh, comet! You have turned evangelist
At – let me see – yes, twenty past eleven,
For even the most cynic atheist
Looks up to heaven.

John O'Keefe, New York *World*

Gee Whyzygy!
As a high old syzygy,
Didn't I throw a scare
Into everybody everywhere?

And didn't I
Make more people look toward the sky
Than anything that has come their way
In many a day? . . .
When they were getting ready to pray,
And turning pale
At thought of my fatal tail,
I swished by
With never a mark on the sky
Or a visible sign
Along the whole starry line –
Not even a smell
Of gas to tell
That I
Was anywhere in the sky . . .

From 'The Comet's Comments' by William J. Lampton, New York *World*

Stranger with fiery mane,
Flame eyes and jewel'd train,
Out of the sun's domain,
Why com'st thou?

Waving a blazing brand,
High in thy far-flung hand;
Thy hair a yellow strand,
Swept from thy brow!

Art thou some fiendish star,
Leaving calm fields afar,
Seeking fresh worlds to mar?
Or, blown astray,

Striving to find a spot,
Ancient and half-forgot,
There, in some ordered lot,
Thy head to lay.

Stephen Chalmers, *New York Times*

Tim O'Mara was conversing with his old pal, Mike Muldoon,
About the Halley comet while they gazed upon the moon.
'A comet's sure a wonder,' said O'Mara looking wise,
'It is,' Muldoon assented, 'an' the greatest in th' skies!

'An' think of all th' books an' things the high-brow fellys write
About the strange beoggerfee of that celestial sight!'
Said O'Mara: 'For live writin's stuff th' comet'll never fail,
And think of all it is because for thereon hangs a tail.'

James Ravenscroft, New York *American*

The astronomers now tell us there's a comet in the sky
Which will quite soon be apparent to the nude observant eye.
But here's a question we, in all humility, propose –
Will the coming comet leave us *comme il faut* or comatose?

N. P. Babcock, New York *American*

Here we are, friends, whole and hale,
In or through the comet's tail;
And, as far as we can say,
Matters are about what they
Were before.

Nothing's added to the stock;
Same old shiver, same old shock.
Round about the sun we'll go
In the same old status quo.
Awful bore!

Chicago *Daily Tribune*

I ain't a-skeered o' comets
A chasin' throo de sky;
I's bein' kep' too busy
Wif de trouble closer by.
De trolley cars is rushin'
An' threatenin' bad luck,
An' as you dodge a motor car
You runs into a truck.

Philander Johnson, Washington *Star*

Is this thy picture? Though no mortal eye
Has seen thy form for many a changing year,
This plate, too sensitive to pass thee by,
Records in black and white thy presence here,
And at thy perihelion in May
The world may hope to see a brave display.

When William lorded it on Hastings field
Over the stricken Saxon, thou wert there;
Nor have the intervening years revealed
The like of that regrettable affair
When Harold's England humbly bowed the knee
To an invading host from o'er the sea.

Thou hast made many passing calls since then,
Nor ever seen upon our island shore
A raider landed, and, as Englishmen,
We bid thee hearty welcome here once more,

Unfolding to our guest from realms unknown
A tale no whit less glorious than his own!

Touchstone, London *Daily Mail* (on reading that the Comet had been successfully photographed at Greenwich)

Out of the dim, the unknown trackless ways,
The comet vagabonds its lawless beam,
As some wild creature to the fallow strays,
With alien tread, to mock the winter's dream.

A fearsome thing that presages of war,
And gaping women turn away the eye;
Rome fell, men say in tones of frightened awe,
As one of these lit up the darkened sky.

The Holy City's walls were battered down
When overhead a blazing comet hung,
And Alexander added crown to crown
Beneath a trail of fire that dipped and swung.

Today we look beyond the Morning Star:
God grant red Carnage may not be unbound!
For, dragon-like, a comet blazes far,
And nations tread with fear on neutral ground.

Charles Henry Chesley, Buffalo *Sunday Times*

His head is broader than the sphere
On which we mortals dwell,
And through the shuddering atmosphere
He rushes on pell-mell;
The planets shriek, the mountains melt,
His path – the sun shrinks from it.
Do we refer to T. Roosevelt?
No, no! to Halley's Comet!

But don't be scared, O timid folk,
Let courage not desert you.
Though big and fierce, he's but a joke,
He has no power to hurt you.
Smile as his shape across our brow
He sweeps his mad waltz – oh!
We speak of Halley's Comet now
And of T. Roosevelt also!

Paul West, New York *World*

Erratic protégé of famous Halley,
What time you swam into our mortal ken,
The frightened folk could not their spirits rally,
Fearing cyanogen!

Cyanogen! Why there is nothing sweeter,
I'd like to live upon it all the time –
You see I've got the gas now in my metre,
And likewise in my rhyme.

But in New York, the time you made your flight in
A way which compromised our poor old earth,
The reckless jokers spent the night in
High revelry and mirth.

Which makes me think: As onwards you went dashing,
If those contrasted scenes you chanced to see,
That thought of Puck's into your head came flashing –
'What fools these mortals be!'

FJC, London *Sunday Times*

(The American Professor Booth had said that if the cyanogen gas in the tail of the Comet united with the hydrogen of earth, all life on this planet would be snuffed out. When the Comet was at its nearest point to earth, 'cyanogen cocktails' were served at New York supper parties.)

The 1985–6 coming of the Comet has so far excited no poetic or versifying reaction, although this situation may change as prime time approaches, especially as there are new elements involved, notably the probes going up to examine the body at close quarters in every detail, and to analyse that mysterious tail. One of our foremost living poets, Seamus Heaney, has already referred to Halley in passing, in *Exposure*, when he finishes with 'The once-in-a-lifetime portent, the comet's pulsing rose.'

Halley's Comet Society, of which the author of this book is the founder, can perhaps be forgiven for adding to the doggerel when welcoming the Managing Director of British Aerospace, Admiral Sir Raymond Lygo, as a member. These were the lines written in the spirit of Philander Johnson, Paul West *et al*!

O Son of Phoebus and Poseidon too,
Join if you will old Halley's faithful crew
And probe the tailpiece of his tadpole guest.
May your swift messenger fulfil your quest
So that in after years *this* Club may raise its hail:
'The body still is Halley's – but Lygo's is the tale!'

Talking of that somewhat exclusive Halley's Comet Society, the fact that Edmond Halley now belongs to the world is underlined by the formation of branches in other parts of the globe, notably in Japan where so much is being done to study and identify the being and implications of the Comet. It is fitting, then, that this brief dissertation on associated verse should conclude with a most moving poem (*true* poetry) by Katsumi Tanaka, as translated by Takamichi Ninomiya and D. J. Enright:

Halley's Comet appeared in 1910
(And I was born in the following year):
Its period being seventy-six years and seven days,
It is due to reappear in 1986 –

So I read and my heart sinks.
It is unlikely that I shall ever see the star –
And probably the case is the same with human encounters.

An understanding mind one meets as seldom,
And an undistracted love one wins as rarely –
I know that my true friend will appear after my death,
And my sweetheart died before I was born.

15

What's What and Who's Who: an ABC Guide to Astronomers

APOLLONIOS (*c.*262–*c.*190 BC). Greek mathematician of Perga. He made important contributions to geometry and named the ELLIPSE, the PARABOLA and the HYPERBOLA. It was his work, the *Conics*, that Halley translated, edited and completed when he became Savilian Professor at Oxford. Apollonios is also supposed to have preceded PTOLEMY in introducing the EPICYCLE to account for the apparently erratic paths of the planets.

ARISTARCHOS (latter half of third century BC). Greek astronomer of Samos. He was able to work out that the sun was bigger than the earth. This was important because it made him come to the conclusion that the earth must revolve around the sun and not vice versa. However, the powerful influence of ARISTOTLE made sure that the GEOCENTRIC view of the universe prevailed until the time of COPERNICUS, when Aristarchos was vindicated.

ARISTOTLE (384–322 BC). Major Greek philosopher whose ideas had a powerful influence on science and thought until the late Middle Ages.

Aristotle wrote on comets in his book the *Meteorologica*. He thought that comets were not actual celestial bodies moving in paths amongst the planets, but rather that they were formed as a result of the upper part of the earth's atmosphere, which he thought was very dry and hot, suddenly catching fire. It was the observational evidence of the comet of 1577 provided by Tycho BRAHE that proved that Aristotle was wrong and that comets did exist beyond the earth's atmosphere.

BRAHE, Tycho (1546–1601). Danish astronomer. In the period just before the telescope was invented, Tycho made careful observation of the planets and so helped to pave the way for KEPLER to formulate his revolutionary three laws of planetary motion.

Tycho shook the astronomical world when he showed with his observations that the SUPERNOVA which exploded in 1572 lay in the region of the stars, i.e. the region that had always been held to be unchanging. In addition, he showed that the bright comet of 1577 moved beyond the earth's atmosphere amongst the planets. This caused a great stir because it meant that, to do this, the comet would have to have smashed through the solid crystal spheres to which the planets were thought to be fixed.

With this discovery, Tycho began the erosion of the belief in the crystal spheres that had existed since ancient times. But he did not support the HELIOCENTRIC universe put forward by COPERNICUS. He still thought that the sun went round the stationary earth, but he now thought that the planets went round the sun and not the earth.

Tycho, who was rather a tyrannical character, built himself a castle and observatory on the island of Ven between Sweden and Denmark. Here he set about making the most accurate astronomical observations on record up till that time without the help of a telescope.

Losing the favour of the new Danish king in 1588, Tycho moved to near Prague where he continued his astronomical work, assisted in 1601 by Kepler, who was then twenty-nine years old.

In 1601 Tycho died and Kepler used his invaluable work to formulate his laws of planetary motion and so substantiate the theory of Copernicus.

It was a great compliment to Halley that FLAMSTEED named him the 'Southern Tycho' after Halley had returned from St Helena where he had charted the stars of the southern hemisphere.

BRUNO, Giordano (1548–1600). Italian thinker who started off life as a Dominican monk. In 1600 he was burned at the stake for his philosophical views that were deemed to be heretical by the Catholic Church.

Bruno was important because he was one of the first men on the continent to be openly sympathetic with the HELIOCENTRIC theory of COPERNICUS. He was a staunch opponent of all dogmatism and held that any view of the universe must necessarily be conditioned by time and space and so cannot be absolutely true.

Although Bruno was not burned as a direct consequence of his Copernican sympathies, his death was ample proof of how religious orthodoxy could react to scientific theories that threatened its authority.

CASSINI, Jean Dominique (in Italian, Giovanni Domenico) (1625–1712). French astronomer of Italian origin. When Louis XIV had built a new observatory at Paris, he invited Cassini to become its first director. Cassini accepted the offer and began working in 1671. Two years later he took French nationality.

In Paris, Halley enjoyed the hospitality of Cassini when he embarked upon his Grand Tour of Europe in 1680. Halley recorded that Cassini thought it possible that the comet that they had both seen recently might have been the reappearance of a previous comet. No doubt Halley bore this in mind when he came to do the calculations that were to result in his successfully predicting 'his own' Comet.

Cassini made important observations of the planet Saturn, discovering a division in its ring and also four of its satellites. After he died, he was succeeded as director of the Paris Observatory by his son, his grandson and his great-grandson.

COPERNICUS, Nicholas (1473–1543) is really Mikoaj Kopernika, Polish astronomer. He revolutionised the PTOLEMAIC view of the universe in which the sun, the moon and the planets went round the earth. Copernicus's HELIOCENTRIC theory, in which the sun is at the centre of the universe, was in fact anticipated nearly 2,000 years before his time by ARISTARCHOS, whose ideas were known to Copernicus.

After studying astronomy at the University of Cracow, Copernicus later became canon of a cathedral in East Prussia. It was his dissatisfaction with the complex system of EPICYCLE and DEFERENT, which was used to explain the motion of planets, that prompted him to reconsider the traditional view of the universe.

Copernicus put forward the theory that the earth revolved around the sun and not vice versa. This he reckoned would make the mathematics involved in accounting for the motion of the planets much easier than before. Indeed, when his book came out in 1543, as Copernicus lay dying, there was a preface in it saying that the contents of the book were just a convenient mathematical fiction and that the earth did not *really* go round the sun. The preface, ostensibly written by a German theologian who felt the need to soften the effect of such radical ideas, was in fact the work of his publisher, Osiander. Certainly, Copernicus himself believed the reality of his own theory.

Copernicus did manage to see a copy of his epoch-making work, called *De Revolutionibus Orbium Coelestium*, before he died. Not only did the book mark the watershed between the classical and medieval astronomical tradition and modern astronomy, it also radically altered man's view of his own importance, since the earth no longer held the central place in the universe but was relegated to the status of an ordinary planet orbiting the sun.

FLAMSTEED, John (1646–1719). English astronomer. Flamsteed was appointed to be the first Astronomer Royal by Charles II in March 1674. Two years later he moved to the Royal Observatory that had just been built at Greenwich, though he had to buy his own instruments and supplement his small salary by teaching privately.

Flamsteed was a moody, bad-tempered man who suffered from ill-health. His hostility towards Halley lasted some thirty years and was probably due, in the last resort, to his envying Halley's talent as an astronomer and his social charm.

Ironically, it was Halley who was given the task by the Royal Society of editing and completing Flamsteed's star catalogue because Flamsteed was reluctant to do it himself. Flamsteed later managed to destroy almost all of the copies of what he saw as a bastard edition of his work. He then set about bringing out his star catalogue himself. It was published posthumously as the *Historia Coelestis Britannica*. In it were charted 3,000 stars and it immediately replaced Tycho BRAHE's catalogue, which it far surpassed in size and accuracy.

GALILEO, Galilei (1564–1642). Italian mathematician and scientist. Galileo was extremely versatile: he investigated the motion of the clock pendulum, he lectured on the centre of gravity in solid bodies and built a telescope through which he made astonishing discoveries.

In 1609, prompted by a new magnifying instrument that had been made in Holland, Galileo set about constructing what was to be the first astronomical telescope. Looking through it, he found to his astonishment that the night sky was different from what traditionally had been thought: he saw the myriad stars of the MILKY WAY, the four satellites of Jupiter and the bumpy surface of the moon. His observations of the moon in particular led him to support the theory of COPERNICUS. Instead of seeing the perfect, smooth gleaming sphere that the moon was supposed to be, he saw a cratered, bumpy surface that was not much different from the earth's.

In 1616, the Copernican theory was denounced by the Catholic Church and Galileo was warned not to support or teach it. Nevertheless, in 1632 he published his famous fictitious dialogue between the supporters of the old Ptolemaic world view and the new Copernican one. Galileo's support of the latter was easily detected and in 1633 he was summoned to Rome to face the Inquisition. Knowing what dreadful fate awaited him if he refused to do what he was told, Galileo renounced his belief and support of the theory of Copernicus.

For the rest of his life, Galileo lived just outside Florence under virtual house arrest. Although hampered by old age and blindness, he managed to write a book on physics which was smuggled out to Holland and published in 1638.

Galileo died in 1642, the year Sir Isaac NEWTON was born. He did much to pave the way for the exploration of the heavens by telescope and he laid the foundations for modern experimental science.

HEVELIUS, Johannes (1611–87). German astronomer born at Danzig (modern-day Gdansk in Poland). He was most famous for his maps of the moon that were published in his work *Selenographia* in 1647. He also produced a substantial star catalogue of 1,564 stars that was published three years after his death.

Hevelius set up an observatory at Danzig that was, when Halley visited him in 1679, the greatest one in Europe, if not the world. Hevelius built his own instruments that included telescopes of up to 150 feet in length.

A controversy arose after 1673 between Hevelius and Robert HOOKE over telescopic sights on observational instruments. Hevelius preferred to use open sights, since he felt that telescopic sights were prone to optical effects. Although the magnifying power of telescopic sights clearly makes them superior to open sights in principle, because at this time they were still at an early stage of development, it was understandable that Hevelius stuck to his open sights. In fact, Halley was able to testify to the excellent results that Hevelius was obtaining with his instruments.

Hevelius suffered the tragedy of having his magnificent observatory burn down in 1679 but he had the courage and the will to rebuild it with the help of well-wishers. He and Halley remained good friends until Hevelius angered Halley by making false accusations against him.

HOOKE, Robert (1635–1703). English physicist. Hooke was an extremely gifted scientist: he was a pioneer in microscope work, studying, amongst other things, snowflakes, corks and fossils; he was very good at mechanics and made many improvements in watches, clocks and astronomical instruments; he built the first Gregorian reflecting telescope; his most famous achievement was his work on the elasticity of materials and his name is commemorated in Hooke's Law. He became a Fellow of the Royal Society in 1663 and later became its Secretary from 1677 to 1683.

Hooke suffered from ill health as a youth and seems to have been mean, grudging and argumentative as an adult. He was involved in controversies with his contemporaries, notably with HEVELIUS, and with NEWTON, when Hooke claimed priority in part of Newton's great work, the *Principia*. Halley had to be the middleman in both disputes, using his tact and eloquence to good effect. It was also Halley's association with Hooke that contributed to FLAMSTEED's dislike of Halley. Any friend of Hooke's was certainly an enemy of Flamsteed's.

KEPLER, Johannes (1571–1630). German astronomer and mathematician, most famous for his three laws on planetary motion (see page 173, KEPLER'S LAWS OF PLANETARY MOTION).

Kepler might have entered the Lutheran Church, but instead became a teacher of mathematics. In 1600, at the age of twenty-nine, he went to Prague and worked with Tycho BRAHE. It was the latter's observational material that helped Kepler to arrive at his three laws of planetary motion.

Kepler discovered that the planet Mars did not move in a perfect circular orbit at a uniform speed as was thought then, but that it moved in an ELLIPSE and that the closer it got to the sun, the faster it went. This discovery helped to establish the Copernican world view as against the Ptolemaic one: the elliptical orbit had triumphed against the aesthetically 'perfect' orbit of the Greeks. He then went on to show a mathematical relationship between the mean distance from a planet to the sun and the time taken by that planet to orbit the sun once.

Kepler also wrote on the relationship between the speeds of planets and notes in a musical scale; he wrote on optics and he designed a telescope that became more popular than Galileo's.

NEWTON, Sir Isaac (1642–1727). English scientist, mathematician and astronomer. He is most famous for his great work, known in short as the *Principia*, one of the greatest scientific books ever written.

Newton was born at Woolsthorpe in Lincolnshire. It was to his mother's house there that he returned from Cambridge and stayed during the plague years of 1665 and 1666.

These two years were a period of great mental fertility for Newton: he developed the branch of mathematics known as calculus; he conceived his famous theory of universal gravitation and he made an important investigation into the nature of white light. His experiments with light showed that the various coloured rays that are obtained when light is passed through a prism are a result of separation by refraction and not due to the varying thickness of the glass, as was then supposed.

Although Newton had conceived the theory of a universal gravitation during 1666, it was not until eighteen years later that he was galvanised into gathering all his thoughts on the matter and putting them down on paper. The initial impetus came from Halley, who made a trip to Cambridge to pick Newton's brains on the problem of working out the mechanical proof needed to show the sort of attraction the sun exerted on a planet. Newton immediately recognised the problem and was able to send the proof to Halley a few days after his visit. Halley realised from his visit that Newton had a great number of scientific ideas and also material of enormous importance. He was determined that they should be made public.

At the urging of Halley, Newton set about writing in eighteen months what was to be one of the greatest scientific books.

Though Newton threatened to suppress the third part of his work after a wrangle with HOOKE, the *Principia* was published in 1687. In it, Newton set out his laws of motion (see page 175, NEWTON'S LAWS OF MOTION); he postulated the idea of absolute time and space, he put forward the principle of universal gravitation and he applied his theoretical work to various phenomena such as tides, comets and the motion of the planets. Halley not only urged Newton to write, but he also financed the project and helped with the editorial work; his contribution was invaluable.

Newton also had less well-known interests; for example, he delved deeply into the strange, arcane world of alchemy and he wrote on the mystical Book of Revelation.

Newton was President of the Royal Society until his death in 1727. His work on optics, his design of a reflecting telescope that used mirrors and not lenses, his development of calculus, his law of motion and his law of universal gravitation all had a profound effect on the course of science.

PTOLEMY, Latin name Claudius Ptolemaeus (second century AD). Greek astronomer, mathematician and astronomer of Alexandria. He is most famous for his work, the *Almagest*, a magnificent synthesis of Greek astronomical thought. It was the PTOLEMAIC SYSTEM that dominated astronomical theory for 1,300 years, until the time of COPERNICUS. He also wrote a book on astrology called the *Tetrabiblos*, and a book on geography in which the longitudes and latitudes of certain places are set out.

ROYAL SOCIETY, THE. This is the popular title of 'The Royal Society of London for improving natural knowledge'. The nucleus of the Society had formed as early as 1645, but meetings began in 1660. In 1662, it received its charter from Charles II. It is now the oldest and premier science academy in the United Kingdom and numbers many eminent foreign scientists amongst its Fellows. In Halley's day, the German astronomer HEVELIUS was a Fellow.

Because of its lack of endowments, the Royal Society had financial problems which eased during Sir Isaac NEWTON's period as President (1703–27). It was the Society's lack of money that threatened the delay in the publication of Newton's great work, the *Principia*, but Halley stepped in and financed the project from his own purse. Among its first Fellows were Robert Boyle, the chemist, John Evelyn, the diarist, and Sir Christopher WREN.

SAVILE, Sir Henry (1549–1622). English scholar who founded the Savilian Professorship of Astronomy and Geometry at Oxford. He was one of the most eminent scholars of the Elizabethan period and worked on the Authorised Version of the Bible that was published in 1611.

WALLIS, John (1616–1703). English mathematician who was ordained in 1640. During the English Civil War, he worked for the Puritans as a de-coder of captured Royalist letters. In 1649 he was appointed Savilian Professor of Geometry at Oxford by Oliver Cromwell. Wallis was a brilliant mathematician and was teaching at Oxford when Halley was an undergraduate there. When Wallis died and the Savilian professorship became vacant, Halley applied for the post and was appointed. In his inaugural speech, Halley was generous in his praise for Wallis.

WREN, Sir Christopher (1632–1723). English architect. Wren is best known for designing the new St Paul's Cathedral after the old one burnt down in the Great Fire of London in 1666. He was also a notable scientist and an eminent member of the ROYAL SOCIETY. He was Professor of Astronomy at Gresham College from 1657 to 1661 and then Savilian Professor of Astronomy at Oxford.

16

Glossary of Scientific Terms

ANTITAIL. Projection effects, when the earth crosses the orbital plane of the comet, sometimes make a portion of the comet's tail appear to point towards the sun.

APHELION. The point in the orbit of a celestial body, such as a comet or planet, when it is furthest away from the sun. Compare PERIHELION.

APPARITION. The period of time that a celestial object is visible from earth. The last apparition of Halley's Comet was in 1910.

ASTEROID (also called a 'minor planet'). More probably the pull of Jupiter prevented any large planet from 'firming' in that region. Any of the estimated 100,000 small planets, ranging from roughly sixty miles to less than one mile in diameter, that are almost all found between the orbits of Mars and Jupiter. It was once thought that they may have originated from the fragments of a planet that broke up or from a very small number of large asteroids that collided with each other.

ASTROLOGY. The field of knowledge which proposes that it is possible to gauge the course or character of people's lives by studying the positions of stars and planets in relation to them. In the first century AD, it was as a result of the astrologer Balbillus that Nero decided to ward off the imagined disastrous effects of a comet by having scores of noblemen killed. There is a letter of Halley's that suggests that he did not give astrology much credence at all.

ASTRONOMICAL UNIT. *Abbr.* A.U. A unit of length used in astronomy to measure distances. It is equal to the average distance between the earth and the sun, i.e. about 93 million miles.

ASTRONOMY. See Chapter 8, ASTRONOMY UP TO THE TIME OF HALLEY, pages 85–7.

ATMOSPHERE. The gaseous envelope that surrounds the earth and other celestial bodies.

AURORA. Plural *aurorae.* A natural phenomenon consisting of a display of streams of glowing, often coloured light thought to be due to the interaction of charged particles from the sun with the upper atmosphere of the earth. It is often seen in places as far south as Scotland. In 1716, Halley published two papers on aurorae in which he advanced the current theories as to their nature.

BIG-BANG THEORY. The currently most favoured cosmological theory. It holds that the universe came into being as a result of the explosion of one massive, dense FIREBALL, up to 20 thousand million years ago. According to the theory, the outwardly exploding material would tally with the fact that the universe can be observed to be expanding outwards.

CELESTIAL BODY. A body, such as a star, planet, comet or asteroid, that exists within the realm of the sky.

CELESTIAL EQUATOR. The projection of the earth's equator on to the celestial sphere.

CELESTIAL SPHERE. The imaginary sphere to which the stars are thought of as being fixed and which slowly revolves around the earth. It is used as a convenient fiction in order to record astronomical data, such as the position of a particular celestial body.

COMA. The volume containing gas and dust around the nucleus of the comet which has not yet been swept into the tails by the solar wind and solar radiation pressure.

CONSTELLATION. A collection of stars that seems from the earth to form a usually recognisable pattern. Two of the most well-known constellations in the northern hemisphere are *Ursa Major*, the Great Bear, often called the Plough, and the W-shaped *Cassiopeia*. By international agreement, there are now eighty-eight officially recognised constellations, whereas forty-eight were listed by Ptolemy in the second century AD.

COPERNICAN. Of or pertaining to Copernicus or his theories. See page 162.

CORONA. The outermost part of the sun's atmosphere that can be seen by the naked eye only during a total solar ECLIPSE, when it appears as a halo around the darkened disc of the moon which is eclipsing the sun.

COSMOLOGY. The field of inquiry concerned with the nature of the universe, including how it originated, how it evolved and how it will develop in the future.

DECLINATION. The angular distance north or south of the celestial equator of a star or other celestial body. Declination is the celestial equivalent of latitude on earth.

DEFERENT. See EPICYCLE.

DIRTY SNOWBALL. A phrase used to describe the model of a comet as proposed by the American astronomer F. L. Whipple. It refers to the hypothesis that the nucleus of a comet is solid, is made up of various ices and small particles and is up to a few miles in diameter. Compare FLYING SANDBANK.

DUST TAIL. Solid dust particles (blown off the nucleus of the comet as it SUBLIMATES), responding to solar radiation pressure and their orbital motion, are pushed away from the nucleus. The dust tail is seen because of sunlight scattered by the dust particles.

EARTH. The third planet from the sun, lying between the orbits of Venus and Mars. It is the only planet known to have life and has one natural satellite – the moon. Up to the late sixteenth century, it was generally believed that the earth was the centre of the universe and that the sun and the planets revolved around it.

ECCENTRICITY. The degree to which an orbit deviates from a circular path. The more elliptical an orbit is, the higher its eccentricity is said to be. Most comets have orbits of high eccentricity.

ECLIPTIC. The plane of the earth's orbital path projected on to the celestial sphere and around which the sun appears to trace out its annual path.

ELLIPSE. A closed curve that looks like an elongated circle. It has two points or foci (plural of focus) along its central axis and the closer together these two foci are, the less elongated and the more circle-like the ellipse is. Most comets have highly elliptical orbits, which means that they will keep coming back to the sun unless they disintegrate or are pulled off course. It was Kepler at the beginning of the seventeenth century who discovered that planets travel in elliptical orbits and not circular ones. Compare HYPERBOLA, PARABOLA.

EPICYCLE. In ancient and medieval astronomy, a circle around whose circumference a planet was supposed to trace out its orbit and whose centre was placed on the circumference of a larger circle called the DEFERENT. The deferent had the earth as its centre. This complex system of epicycle and deferent was used to account for the motions of the planets before KEPLER eventually showed that planets have elliptical paths.

EQUINOX. Either of the two occasions during the year at about March 21st (*vernal equinox*) and September 23rd (*autumnal equinox*) when the sun crosses the celestial equator and day and night are of equal duration in all parts of the world. (Equinox means 'equal night'.)

ETHER, THE. A hypothetical transparent substance once thought by scientists, especially during the nineteenth century, to have filled all space.

EUROPEAN SPACE AGENCY. *Abbr.* ESA. An organisation of eleven European countries, including the United Kingdom and Eire, for the furthering of information about space and the development of spacecraft. The ESA plan to send up a SPACE PROBE to find out more about Halley's Comet (see Chapter 1).

FIREBALL. A large, brilliantly luminous meteor, especially one that has a sparkling trail lasting for several seconds or more. It is possible for a fireball to be brighter than the moon.

FLUORESCENCE. The emission of light of a longer wavelength after absorption of shorter wavelength electromagnetic radiation by atoms, molecules, or ions.

FLYING SANDBANK. A phrase used to describe the model of a comet as championed by the British astronomer R. A. Lyttleton. It refers to the hypothesis that the nucleus of a comet is qualitatively the same as the COMA, i.e. both are composed of tiny independent particles that become more dense towards the centre of the coma, giving the false impression of a solid nucleus. The flying sandbank model is currently less favoured than the DIRTY SNOWBALL.

GALAXY. Any of the billions of systems of stars, nebulae, interstellar matter, etc., that exist within the universe. Galaxies have been classified according to their shape into three basic types: ellipticals, spirals and irregulars. The spiral galaxy to which our SOLAR SYSTEM belongs contains some hundred thousand million stars, one of which is our sun. It is disc-shaped and our sun lies about 30,000 LIGHT YEARS from its centre. It is also called the 'Milky Way Galaxy'. Incidentally our Galaxy with a capital 'G' is distinguished from all other galaxies which are given a small 'g'.

GEGENSCHEIN. Literally meaning 'counterglow', this phenomenon of the zodiacal light is due to sunlight back-scattered from interplanetary dust located outside the earth's orbit and opposite the sun in the sky.

GEOCENTRIC. An adjective meaning 'centred around the earth'. In the geocentric view of the universe, as put forward by ARISTOTLE and PTOLEMY, the earth lay at the centre and the sun, moon and planets revolved around it. Compare HELIOCENTRIC.

GREAT CIRCLE. A line on the surface of a sphere obtained by a plane cutting the sphere's centre. On the earth, lines of longitude and the equator are great circles. On the CELESTIAL SPHERE, the CELESTIAL EQUATOR is a great circle.

HEAD. The nucleus and COMA of the comet are collectively referred to as the head.

HELIOCENTRIC. An adjective meaning 'centred around the sun'. In the heliocentric view of the universe, as established by COPERNICUS, the sun lies at the centre with the earth, moon and planets revolving round it. Compare GEOCENTRIC.

HYDROGEN ENVELOPE. Seen only in ultra-violet light, this gigantic cloud of atomic hydrogen surrounds the comet's head.

HYPERBOLA. An open curve that resembles a parabola except that its two arms are not parallel but diverge outwards. If a comet has a hyperbolic path it means that once it has gone round the sun, it will head off into outer space, never to return. Compare ELLIPSE, PARABOLA.

INTERSTELLAR MATTER. The minute particles and gas of relatively low density that exist throughout the GALAXY.

ION, IONISE. An ion is a neutral atom or molecule which acquires additional positive or negative charge. Solar ultra-violet radiation is the principal reason neutrals become ionised in comets.

ION TAIL. The parent molecules released by the nucleus are ionised by sunlight and dragged away by the magnetic field carried by the solar wind to form the ion tail. The tail is seen by the light of fluorescing ions.

JOVIAN PLANETS. The four giant planets Jupiter, Saturn, Uranus and Neptune.

JUPITER. The fifth planet from the sun, lying between the orbits of Mars and Saturn. It is the largest planet in the SOLAR SYSTEM with a volume so huge that 1,319 earths could fit into it. It takes 11.9 years to make one complete orbit. One of its interesting features is its 'Great Red Spot' which is thought to be a giant cyclone. There was once a theory that this 'spot' was a great volcano that pumped out new comets into space. Because of its powerful gravitational field, Jupiter can easily affect the paths of comets.

KEPLER'S LAWS OF PLANETARY MOTION. There are three laws: 1. A planet has an elliptical orbit round the sun which is situated at one of the two foci of the ELLIPSE. 2. The RADIUS VECTOR between the sun and a planet sweeps out equal areas of space in equal times, so that the closer a planet is to the sun the quicker it travels. 3. The square of the time taken by a planet to orbit the sun once is directly in proportion to the cube of the mean distance between the sun and the planet.

It was in connection with the third law that Halley went to consult Newton at Cambridge. This visit consequently led Halley to urge Newton to write down all his ideas into one of the greatest scientific works ever written, the *Principia*.

LIGHT YEAR. A unit of length used for measuring great distances in the universe. It is equal to the distance light travels in one year in a vacuum, i.e. 5.8786 billion miles.

LONG-PERIOD. Designating a comet that takes a relatively long time to complete an orbit, i.e. from over a couple of hundred years to tens of thousands of years. Compare SHORT-PERIOD.

LUNATION. The period of time, on average 29.53 days, taken by one full lunar cycle round the earth, measured between two successive new moons.

MARS. The fourth planet from the sun, lying between the orbits of the earth and Jupiter. Mars has seasons similar to those of earth, only they are longer. Mars takes almost 687 days to complete one orbit and has two small moons.

MERCURY. The planet that lies closest to the sun. Its cratered surface resembles the moon's and it takes only 87.97 days to make one complete orbit. It was the TRANSIT of Mercury across the sun's disc that Halley saw during his stay in the island of St Helena and which he suggested could be used for finding out the distance of the sun from the earth.

MERIDIAN. On the earth's surface, a GREAT CIRCLE that goes through both north and south poles and cuts the equator at right angles. The celestial meridian is the projection of the earth's meridian on to the celestial sphere.

METEOR. Any of a large number of particles, usually smaller than pinheads, that are only observable when they come into contact with the earth's atmosphere at a speed of up to fifty miles per second. They then burn up in what seems to an observer on earth a sudden streak of light, dying in a matter of seconds. Also called 'shooting star'.

METEORITE. A solid, rock-like piece of matter large enough to survive the burning process that happens when it comes into contact with the earth's atmosphere and which reaches the surface of the earth.

METEOROID. Any of the meteoritic bodies that exist in space and that can potentially become meteors or meteorites.

METEOR SHOWER. A high proportion of meteors visible over a short period of time in a particular area of the sky. It is thought that these showers are formed from debris scattered from a comet. For example, the Orionid shower in late October is thought to originate from Halley's Comet.

MILKY WAY. The stretch of faint milky-white light that can be seen in a clear, moonless night sky; it comes from the combined starlight of millions of stars in the GALAXY which contains our own SOLAR SYSTEM. Because of the shape of our Galaxy, the Milky Way represents the part where the stars are lying in the same plane and seem most dense to an observer.

MILKY WAY GALAXY. See GALAXY.

MOON. The only natural satellite of the earth. It is the chief cause of the tides. Halley embarked on an eighteen-year programme of observing the moon's positions which he thought would solve the problem of determining longitude at sea.

NEBULA. A cloud of dust and gases in space that can be seen as either a diffuse patch of light or a region of relative darkness, depending upon the nature and density of the clouds. In a paper of 1715, Halley made a relatively accurate guess as to the nature of nebulae, long before the advent of the SPECTROSCOPE, which opened up the investigation of nebulae.

NEPTUNE. The eighth planet from the sun, lying between the orbits of Uranus and Pluto. The work of Le Verrier, French astronomer, led to the first optical identification of Neptune. Taking 164.8 years to complete one orbit, it can only be seen through a very powerful telescope.

NEWTON'S LAWS OF MOTION. 1. Unless a force acts on it, a body will stay at rest or in uniform motion. 2. If a force acts upon a body, it will accelerate in the direction of and in proportion to the force, and in inverse proportion to its mass. 3. There is an equal and opposite reaction to every action.

NONGRAVITATIONAL FORCES. Forces changing a cometary orbit which are not due to gravitational effects; usually identified with rocket-like forces on the nucleus (the so-called 'rocket effect').

NOVA. A star that suddenly, in a matter of hours or days, becomes intensely bright as a result of an internal explosion. That Halley was interested in novae is shown by a paper published in 1715 in which he studies novae going back to 1550.

NUCLEUS. See COMET.

OCCULTATION. The temporary obscuring of the light of one celestial body by another, as when the moon moves in front of a planet or a star.

OPPOSITION. The moment when one celestial body is opposite or nearly opposite another in the sky. When one of the outer planets, i.e. Mars, Jupiter, Saturn, Uranus, Neptune and Pluto, is at opposition it lies in a straight line with the sun and is at its closest to the earth, which lies between it and the sun.

ORBIT. The usually elliptical path taken by one celestial body, such as a planet or comet in a gravitational field, around another, such as the sun. The orbit of Halley's Comet takes from seventy-six to seventy-nine years to complete, the discrepancy being due to the gravitational effects of Jupiter or another large planet.

PARABOLA. A U-shaped open curve with both its arms parallel to each other and extending out to infinity. If a comet had a parabolic path, it would orbit the sun and then head off into outer space, never to return. Compare ELLIPSE, HYPERBOLA.

PARALLAX. The change in the position of a star or other celestial body when observed from two sufficiently distant positions on the surface of the earth.

PARENT MOLECULES. Water (H_2O), carbon dioxide (CO_2), hydrogen cyanide (HCN) and other molecules containing carbon and sulphur are believed to be the source molecules for many of the neutral and ionised atomic and molecular species observed in the coma and tail of a comet.

PENUMBRA. The outer, partially dark region of the shadow cast by a celestial body, such as the earth or moon, during an eclipse.

PERIHELION. The point in the orbit of a celestial body, such as a comet or planet, when it is nearest to the sun. Compare APHELION.

PERTURBATION. An effect that alters the course of an orbiting celestial body, such as a comet, caused by the gravitational field of a larger celestial body, such as a planet. In calculating the orbit of a comet the perturbations from Jupiter and Saturn usually have to be taken into account.

PLANET. A celestial body that orbits a star whose light it reflects. In our SOLAR SYSTEM there are nine planets which revolve around the star we call the sun.

PLASMA. A 'gas' of positive and negative ions.

PLASMA TAIL. A different name for an ION TAIL.

PLUTO. Ninth and outermost planet of the SOLAR SYSTEM, but at present is closer in than Neptune and will remain so until 1999. Pluto was discovered in 1930 after an extensive search had been made for it. Pluto is so far away that information about it is difficult to obtain. It is suggested that it might even be a former satellite of Neptune whose gravitational field it escaped. It takes 247.7 years to complete one orbit.

POLARIS. The north star or Pole star.

PTOLEMAIC SYSTEM. The view of the universe as put forward by Ptolemy. It supposed that the earth lay at the centre with the moon, Mercury, Venus, the sun, Mars, Jupiter and Saturn all attached to solid, transparent, crystal spheres and moving round the earth. Beyond Saturn lay the stars fixed in place to their crystal sphere. The moon, planets and the sun were thought to orbit in perfect circular paths (the circle being considered the aesthetically perfect figure). However, to account for the observed irregularities of the planets' paths a complicated system of EPICYCLE and DEFERENT was devised which became obsolete only following the astronomical revolution led by Copernicus and Kepler. It was Copernicus who established that the earth went round the sun and Kepler who showed that planets have elliptical orbits and not circular ones.

QUADRANT. An instrument used for measuring the altitude of the stars. It typically consists of a graduated arc of ninety degrees and movable sights.

RADIATION PRESSURE. Electromagnetic radiation (e.g. light, infrared, X-rays, radio, ultra-violet, etc.) has the property of being able to transfer momentum – push – materials away from the source of the radiation.

RADIUS VECTOR. An imaginary straight line, used for the purpose of calculation, that joins one celestial body, such as the sun, with another, such as a comet, that is orbiting it.

RIGHT ASCENSION. The angular distance between the vernal EQUINOX and a particular point on the CELESTIAL EQUATOR. It is measured eastwards from the vernal equinox in units of hours, minutes and seconds. It is the celestial equivalent of longitude on earth.

SAROS. *Adj.* sarotic. A cycle during which eclipses of the moon and the sun occur in a recognisable pattern. The cycle lasts for eighteen years and 11.3 days and is then followed by another very similar though not identical cycle. Halley embarked on a programme of observing the moon through a sarotic period in the hope of solving the problem of determining longitude at sea.

SATELLITE. An object or body that orbits a larger body. For example, the moon is a satellite of the earth.

SATURN. The sixth planet from the sun, lying between the orbits of Jupiter and Uranus. It is famous for its rings which consist of millions of tiny particles that together orbit the planet. Saturn is the second largest planet in the SOLAR SYSTEM and takes 29.5 years to make one complete orbit. It has twenty (possibly more) satellites of which Titan is the largest. Saturn's powerful gravitational field can easily affect the path of a comet.

SCATTERING. Small particles (one micrometer to 1/10 millimeter in size) have the property of not simply reflecting light and making shadows but actually scatter the light that illuminates them in all directions. In some situations forward-scattered light, appearing where a shadow would be expected, is actually brighter than back-scattered ('reflected') light.

SELF-LUMINOUS. Designating a celestial body that has the capacity to emit light. The sun and other stars are self-luminous but the earth and the other planets are not.

SEXTANT. A measuring instrument consisting of a small telescope, a graduated arc of sixty degrees and two mirrors, one of which is movable. It is especially used in navigation to ascertain the altitude of a particular celestial body above the horizon.

SHOOTING STAR. Popular name for METEORITE.

SHORT-PERIOD. Designating a comet that takes a relatively short time to complete an orbit, i.e. from a couple of years to a couple of hundred years. Compare LONG-PERIOD.

SOLAR SYSTEM. The system within the MILKY WAY GALAXY in which the nine planets – Mercury, Venus, Earth, Mars, Jupiter, Saturn, Uranus, Neptune and Pluto – their satellites, the comets, the asteroids and the meteoroids all orbit the sun.

SOLAR WIND. The continuous stream of atomic particles ejected at a great speed (up to 500 miles per second) from the sun into the SOLAR SYSTEM. The interaction of the solar wind with a comet's tail causes the latter to point away from the sun.

SPACE PROBE. An unmanned space vehicle sent to gather various astronomical information, such as the nature of a planet's surface and atmosphere.

SPECTROSCOPY. The study and analysis of the spectrum of a particular source of radiation, such as a star or a comet. By means of spectroscopy, it is possible to learn certain things about a celestial body, such as its chemical composition and temperature.

SPECTRUM. *Plural* spectra. Broadly, the breakdown of electromagnetic radiation into its constituent wavelengths. When white light is passed through a prism, the band of seven colours, ranging from red to violet, which is produced is the spectrum. The spectra that come from sources where radiation has been emitted are called *emission spectra.* The spectra that come from sources where radiation has been absorbed are called *absorption spectra.*

STAR. A SELF-LUMINOUS celestial body composed of gas that produces heat and light as a result of internal nuclear reactions. Our sun is a star.

STEADY-STATE THEORY. A cosmological theory that puts forward the hypothesis that the universe has always existed, is infinite in time and space and that, as the galaxies evolve and recede outwards, new galaxies come into being so that the overall average density of the UNIVERSE does not alter. In recent times, this theory has become less popular than the BIG-BANG THEORY.

STRIAE. Narrow, rectilinear structures sometimes seen in the DUST TAIL. They are made of particles that were released at the same time from the NUCLEUS and later disintegrate into fragments.

SUBLIMATE. The change of state directly from solid to gas without going through a liquid phase.

SUN. The star that is at the centre of our SOLAR SYSTEM and around which revolve the nine planets, the comets, the meteoroids and other celestial bodies. There are some 100 thousand million other stars in our GALAXY.

SUNSPOT. Any of the relatively cool, dark spots on the surface of the sun that can usually only be observed with a telescope. A sunspot can range up to many thousands of miles in diameter and possesses a strong magnetic field. When Halley was a student at Oxford in 1676, he made observations on sunspots and these were published in *Philosophical Transactions*.

SUPERNOVA. A star that suddenly increases its luminosity by up to 100 million times as a result of a catastrophic internal explosion.

SYNCHRONES. The loci of particles released from the NUCLEUS simultaneously. They are sometimes seen in the DUST TAIL as straight or moderately curved structures.

SYNDYNAMES. The loci of particles in the DUST TAIL that are subjected to equal force.

TAIL. The part of a comet that is composed of dust particles and gaseous material and streams away from the COMA. A tail can extend to up to tens of millions of miles but not all comets have tails. The tail always points away from the sun because of its interaction with the SOLAR WIND.

TRANSIT. The passage of a smaller celestial body, such as Mercury, across the disc of a larger one, such as the sun. Halley proposed that the distance of the sun from the earth could be calculated from observations made of the transit of Mercury or Venus across the sun.

UNIVERSE. The universe is the sum total of all that exists. It incorporates the billions of galaxies that in turn incorporate billions of stars.

URANUS. The seventh planet from the sun, lying between the orbits of Saturn and Neptune. It was discovered in 1781 by Sir William Herschel. It has a diameter roughly four times that of earth. It takes 84.01 years to make one orbit and has five known moons.

VENUS. The second planet from the sun, lying between the orbits of Mercury and the earth. It has a very hot and rocky surface and takes 224.7 days to make one complete orbit. In 1679, Halley recommended the TRANSITS of Venus across the disc of the sun to be used for finding out the distance of the sun from the earth. Venus is also known as the 'morning star' and the 'evening star'.

ZODIACAL BAND. The faint glow seen along the ECLIPTIC connecting the ZODIACAL LIGHT PYRAMIDS to the GEGENSCHEIN.

ZODIACAL LIGHT. A general glow throughout the sky caused by sunlight scattered by interplanetary dust. It is brightest near the sun and along the ECLIPTIC. The ZODIACAL LIGHT PYRAMIDS are often referred to as the zodiacal light.

ZODIACAL LIGHT PYRAMID. This triangular glow seen on the western horizon after evening twilight and on the eastern horizon before morning twilight is the brightest component of the ZODIACAL LIGHT.

Index

HOW TO FIND HALLEY'S COMET –

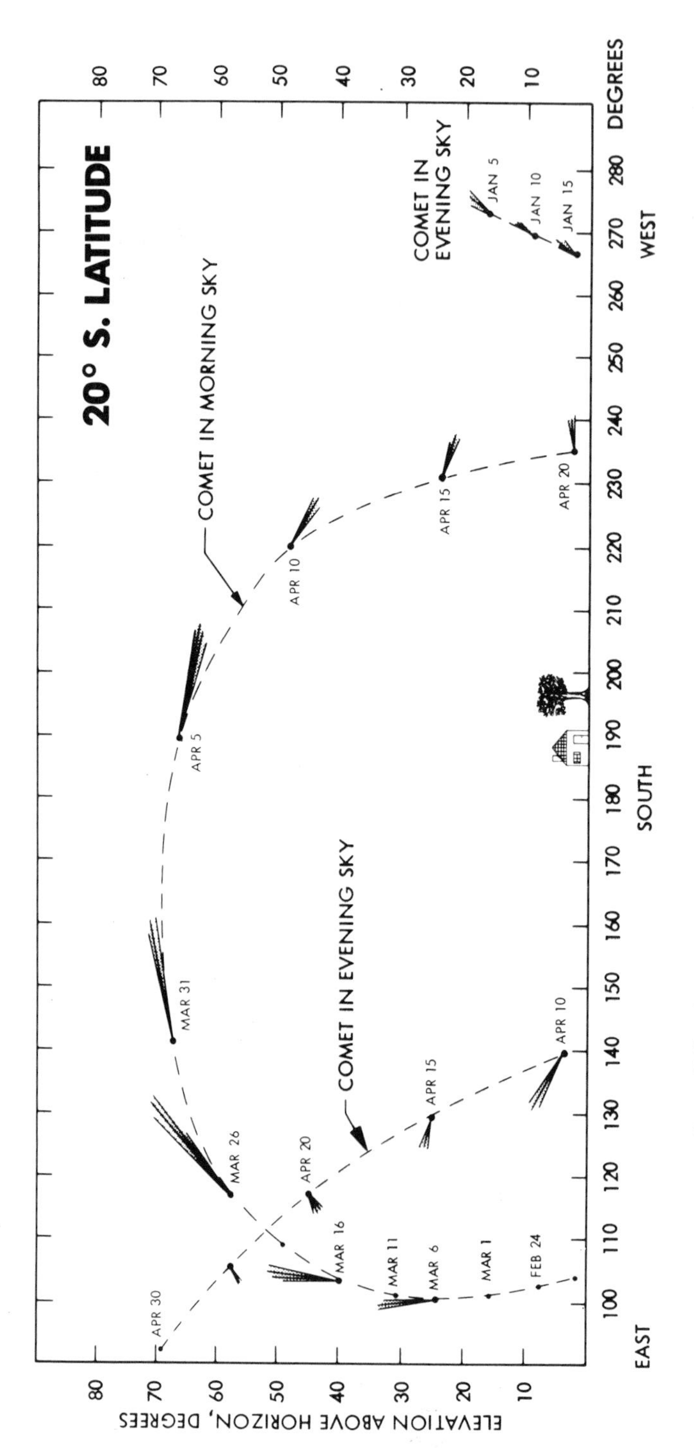

Comet Halley observing conditions in 1986 for observers located at 20° South Latitude (for appropriate countries see Harpur's Guide pages 36 and 37). Comet positions are given for beginning of morning astronomical twilight or end of evening astronomical twilight. Viewing with binoculars and ideal observing conditions are assumed. Astronomical twilight is approximately one hour before Sun rises or sets. For example on 10th April one observes Comet is high in morning sky looking South West.